RICHARD NISBETT

Mindware

Tools for Smart Thinking

PENGUIN BOOKS

PENGUIN BOOKS

UK | USA | Canada | Ireland | Australia
India | New Zealand | South Africa

Penguin Books is part of the Penguin Random House group of companies
whose addresses can be found at global.penguinrandomhouse.com.

First published in the United States of America by Farrar, Straus and Giroux 2015
First published in Great Britain by Allen Lane 2015
Published in Penguin Books 2016
001

Grateful acknowledgment is made to Roger Shepard for permission to print the
image on page 17, and to Takahiko Masuda for permission
to print the images on pages 46 and 47.

Set in 9.79/12.67 pt Electra LT Std
Typeset by Jouve (UK), Milton Keynes
Printed in Great Britain by Clays Ltd, St Ives plc

A CIP catalogue record for this book is available from the British Library

ISBN: 978–0–141–97627–3

www.greenpenguin.co.uk

Penguin Random House is committed to a
sustainable future for our business, our readers
and our planet. This book is made from Forest
Stewardship Council® certified paper.

For Sarah Nisbett

Contents

Conclusion: The Tools of the Lay Scientist 275

MINDWARE

Introduction

The logic of science is the logic of business and of life.

—John Stuart Mill

In an earlier era, when many people were involved in surveying land, it made sense to require that almost every student entering a top college know something of trigonometry. Today, a basic grounding in probability, statistics and decision analysis makes far more sense.

—Lawrence Summers, former president of Harvard University

The word "cosine" never *ever* comes up.

—Roz Chast, *Secrets of Adulthood*

You paid twelve dollars for a ticket to a movie that, you realize after a half an hour of watching it, is quite uninteresting and tedious. Should you stay at the theater or leave?

You own two stocks, one of which has done quite well over the past few years and the other of which has actually lost a little since you bought it. You need some money and have to sell one of the stocks. Do you sell the successful stock in order to avoid locking in your losses on the unsuccessful stock, or do you sell the unsuccessful stock in hopes that the successful one will continue to make more money?

You must choose between two candidates for a job. Candidate A has more experience and stronger recommendations than candidate B, but in the interview candidate B seems brighter and more energetic than candidate A. Which do you hire?

You are a human relations officer for a company. Several women have written you to complain that their job applications were turned down in favor of men with poorer job qualifications. How could you go about finding whether there is really discrimination on the basis of gender?

Time magazine recently reported that parents should not try to control their children's food intake because parents who do that have children who are likely to be overweight. Do you see anything dubious about that claim?

People who have a drink or two of alcohol once a day have fewer cardiovascular problems than people who don't. Assuming you drink less than that, should you increase your alcohol intake? Should you decrease it if you drink more than that?

Problems like those above don't appear on IQ tests, but there are smarter and less smart ways of solving them. By the time you've finished this book, you'll have a cognitive tool kit that will allow you to think about such problems—and infinitely more besides—much differently than you would now. The tools are one hundred or so concepts, principles, and rules of inference developed by scientists in many fields—especially psychology and economics—and by statisticians, logicians, and philosophers. Sometimes commonsense approaches to problems produce errors in judgment and unfortunate actions. The concepts in this book will show you how to think and act more effectively. The ideas provide a supplement to common sense—rules and principles you can learn to apply automatically and effortlessly to countless problems that crop up in everyday life.

This book addresses some of the most fundamental questions about how to reason and make valid inferences. What counts as an explanation (for everything from why our friend acts in such an annoying way to why a product launch failed)? How can we tell the difference between events that are causally related and events that are merely associated with each other in time or place? What kinds of knowledge can be considered certain and what kinds only conjectural? What are the characteristics of a good theory—in science and in everyday life? How can we tell the difference between theories that can be falsified and those that can't? If we have a theory about what kinds of business or professional practices are effective, how can we test that theory in a convincing way?

The media bombard us with alleged scientific findings, many of which are simply wrong. How can we evaluate conflicting scientific claims we encounter in the media? When should we trust the experts—assuming we can find them—and when should we be dubious?

And most important, how can we increase the likelihood that the choices we make will best serve our purposes and improve the lives of ourselves and others?

Can Reasoning Really Be Taught?

But can people actually be taught to think more effectively? Not just to know more things, such as the capital of Uzbekistan or the procedure for extracting square roots, but actually to reason more correctly and solve personal and professional problems more satisfactorily.

The answer to this question is far from obvious, though for twenty-six hundred years many philosophers and educators were confident that reasoning could be taught. Plato said, "Even the dull, if they have had arithmetical training, . . . always become much quicker than they would otherwise have been . . . We must endeavor to persuade those who are to be the principal men of our state to go and learn arithmetic." Later, Roman philosophers added studying grammar and exercising the memory to the practices that would improve reasoning. The medieval scholastics emphasized logic, particularly syllogisms (e.g., All men are mortal. Socrates is a man, therefore Socrates is mortal). The humanists of the Renaissance added Latin and Greek, possibly because they thought that using those languages contributed to the success of these ancient civilizations.

The faith in drilling mathematical, logical, and linguistic rules was strong enough that by the nineteenth century some people believed that pure exercise of the brain on difficult rule systems—*any* difficult rule system—was enough to make people smarter. A nineteenth-century educator was able to maintain, "My claim for Latin, as an Englishman and a teacher, is simply that it would be impossible to devise for English boys a better teaching instrument. The acquisition of a language is educationally of no importance; what is important is the process of acquiring it. The one great merit of Latin as a teaching instrument is its tremendous difficulty."

There never was a shred of evidence for any of these educational views—from those of Plato to those of the fusty old Latin teacher. So psychologists in the early twentieth century set out to produce some scientific evidence about reasoning and how to improve it.

The early returns were not good for what had come to be called "formal discipline"—training in how to think as opposed to what to know. At the turn of the century, Edward Thorndike maintained that no amount of brain exercise or drilling in abstract rules of thought would serve to make people smarter and pronounced the "learning Latin" theory of education to be defunct. He claimed that his experiments showed that "transfer of training" from one cognitive task to another occurred only if the problems were extremely similar in their concrete features. But the tasks that Thorndike studied didn't really qualify as ones involving reasoning: for example, he found that practicing cancellation of letters in a sentence produced no increase in speed of canceling parts of speech in a paragraph. This is scarcely what you would think of as reasoning.

Herbert Simon and Allen Newell, the great midcentury computer scientists, also claimed that people couldn't learn abstract rules for reasoning and provided somewhat better evidence. But their argument was based on very limited observations. Learning how to solve the Towers of Hanoi problem (transferring a stack of disks from one pole to another without ever placing a bigger disk on a smaller one, a game you may have played as a child) didn't produce improvement on the Missionaries and Cannibals problem, which requires a plan for getting missionaries across a river without ever allowing the missionaries on the boat to be out-numbered by the cannibals. The two problems have the same formal structure, but there was no transfer of training on how to solve one prob-lem to ability to solve the other. This result was interesting, but scarcely enough to convince us that training on a given problem could never generalize to solving a problem with a similar structure.

Jean Piaget, the great Swiss cognitive psychologist who studied chil-dren's learning, was an exception to the mid-twentieth-century consen-sus against abstract rules for reasoning. He believed that people did indeed possess such rules, including logical rules and "schemas" for understanding concepts such as probability. But he believed such rules couldn't be taught; rather, they could only be induced as the child encounters more and more problems that can be solved using a particular rule that she discovers for herself. Moreover, the set of abstract rules for understanding the world is complete by adolescence, and every cognitively normal person ends up with the exact same set of rules.

Piaget got it right about the existence of abstract concepts and rule systems that people could apply in everyday life but got everything else wrong. Such rule systems can be taught as well as induced—we keep on learning them well beyond adolescence—and people differ dramatically in the particular set of abstract rules for reasoning that they use.

The early twentieth-century psychologists opposed to the concept of formal discipline were right about one very important matter: getting smarter is not a matter of sheer exercise of the brain. The mind is like a muscle in some ways but not in others. Lifting pretty much anything will make you stronger. But thinking about just anything in any old way is not likely to make you smarter. Learning Latin almost certainly produces little gain in reasoning ability. The nature of the concepts and rules you're trying to learn is everything when it comes to building the muscles of the mind. Some are useless for building brain muscles and some are priceless.

Ideas That Travel

The idea for this book came from my fascination with the fact that scientists' ideas in one field can be extremely valuable for other fields. A favorite buzzword in academia is "interdisciplinary." I'm pretty sure that some people who use the word might not be able to tell you why interdisciplinary research is a good idea. But it is, and here's why.

Science is often described as a "seamless web." What's meant by that is that the facts, methods, theories, and rules of inference discovered in one field can be helpful for other fields. And philosophy and logic can affect reasoning in literally every field of science.

Field theory in physics gave rise to field theory in psychology. Particle physicists use statistics developed for psychologists. Scientists studying agricultural practice invented statistical tools that are crucial for behavioral scientists. Theories developed by psychologists to describe how rats learn to run mazes guided computer scientists in their effort to teach machines how to learn.

Darwin's theory of natural selection owes a great deal to eighteenth-century Scottish philosophers' theories about social systems, in particular Adam Smith's theory that societal wealth is created by rational actors pursuing their own selfish interests.[1]

Economists are now making major contributions to the understanding of human intelligence and self-control. Economists' views about how people make choices were transformed by cognitive psychologists, and economists' scientific tools were greatly expanded by adopting the experimental techniques used by social psychologists.

Modern sociologists owe a great deal to eighteenth- and nineteenth-century philosophers who theorized about the nature of society. Cognitive psychologists and social psychologists are broadening the range of questions raised by philosophers and have begun to propose answers to some long-standing philosophical conundrums. Philosophical questions about ethics and theory of knowledge guide the research of psychologists and economists. Neuroscience research and concepts are transforming psychology, economics, and even philosophy.

A few examples from my own research will show how extensive the borrowing can be from one field of science to another.

I was trained as a social psychologist, but most of my early research dealt with feeding behavior and obesity. When I began the work, the lay assumption, as well as the scientific and medical view, was that people who are overweight get that way by eating too much. But eventually it became clear that most overweight people were actually hungry. Psychologists studying obesity borrowed from biology the homeostatic concept of a "set point." The body attempts to maintain a set point for temperature, for example. The obese have a set point for the ratio of fat to other tissue that differs from that of normal-weight people. But social norms drive them toward being thin, with the result that they're chronically hungry.[2]

The next problem I studied was how people understand the causes of the behavior of other people and themselves. Field theory in physics prompted research showing that situational and contextual factors are often more important in producing behavior than personal dispositions such as traits, abilities, and preferences. This conceptualization made it easy to see that our causal explanations for behavior—our own, other people's, and even that of objects—tend to slight situational factors while overemphasizing dispositional factors.

In studying causal attributions it became clear to me that much of the time we have very limited insight into the causes of our own behavior; and we have no direct access at all to our thought processes. This work on self-awareness owes a great deal to Michael Polanyi, a chemist turned philosopher of science.[3] He argued that much of our knowledge, even about matters we deal with in our field of expertise—perhaps especially about such matters—is "tacit" and difficult or impossible to articulate. Work by me and others on the vagaries of introspection called into question all research that depends on self-reports about mental processes and the causes of one's own behavior. Measurement techniques in psychology and throughout the behavioral and social sciences have changed as a result of this work. The research has also convinced some students of the law that self-reports about motives and goals can be highly unreliable—not for reasons of self-enhancement or self-protection, but because so much of mental life is inaccessible.

The errors discovered in self-reports led me to a concern with the accuracy of our inferences in everyday life in general. Following the cognitive psychologists Amos Tversky and Daniel Kahneman, I compared people's reasoning to scientific, statistical, and logical standards and found large classes of judgments to be systematically mistaken. Inferences frequently violate principles of statistics, economics, logic, and basic scientific methodology. Work by psychologists on these questions has influenced philosophers, economists, and policy makers.

Finally, I've done research showing that East Asians and Westerners sometimes make inferences about the world in fundamentally different ways. This research was guided by the ideas of philosophers, historians, and anthropologists. I became convinced that Eastern habits of thought, which have been called dialectical, provide powerful tools for thinking that can benefit Westerners as much as they have helped Easterners for thousands of years.[4]

Scientific and Philosophical Thinking Can Be Taught in Ways That Affect Reasoning in Everyday Life

My research on reasoning has had big effects on my reasoning in everyday life. I am constantly discovering that many of the concepts that travel across scientific fields are also affecting my approach to professional and personal problems. At the same time, I am constantly being made aware of my failures to use the kinds of reasoning tools I study and teach.

Naturally I began to wonder whether other people's thinking about everyday life events are affected by training in concepts learned in school. Initially I was quite dubious that a course or two dealing with one or another approach to reasoning could have the kind of impact on people that long exposure to the concepts had on me. The twentieth-century skepticism about the possibility of teaching reasoning continued to influence my thinking.

I could not have been more mistaken. It turns out that the courses people take in college really do affect inferences about the world—often very markedly. Rules of logic, statistical principles such as the law of large numbers and regression to the mean, principles of scientific methodology such as how to establish control groups when making assertions about cause and effect, classical economic principles, and decision theory concepts all influence the way people think about problems that crop up in everyday life.[5] They affect how people reason about athletic events, what procedures they think are best for going about the process of hiring someone, and even their approach to such minor questions as whether they should finish a meal that isn't very tasty.

Since some university courses greatly improve people's reasoning about everyday life events, I decided to see whether I could teach such concepts in the laboratory.[6] My coworkers and I developed techniques for teaching inferential rules that are helpful for reasoning about common personal and professional problems. As it turned out, people readily learned from these brief sessions. Teaching about the statistical concept of the law of large numbers affected their reasoning about how much evidence is needed to reach accurate beliefs about some object or person. Teaching about the economic principle of avoiding opportunity costs affected how they reasoned about time usage. Most impressively, we sometimes questioned subjects weeks later, in contexts where they didn't know they were being studied, such as telephone polls allegedly being conducted by survey researchers. We were delighted to find that people often retained substantial ability to apply the concepts to ordinary problems outside the laboratory context in which the concepts had been taught.

Most important, we discovered how to greatly extend the reach of inferential rules to problems of everyday life. We can have complete command over good principles for reasoning in a particular field and yet be unable to apply them to the full panoply of problems that we encounter in everyday life. But these inferential principles can be made more accessible and usable. The key is learning how to *frame* events in such a way that the relevance of the principles to the solutions of particular problems is made clear, and learning how to *code* events in such a way that the principles can actually be applied to the events. We don't normally think of forming impressions of an individual's personality as a statistical process consisting of sampling a population of events, but they are exactly that. And framing them in that way makes us both more cautious about some kinds of personality ascriptions and better able to predict the individual's behavior in the future.

There were several criteria that guided me in selecting particular concepts to write about.

1. The concept had to be important—for science and for life. There are scores of syllogisms that have been around since the Middle Ages, but only a few have even the remotest relevance to everyday life, and it's those that are in the book. There are hundreds of types of fallacious reasoning that have been identified, but only a relative few are mistakes that intelligent people actually make with any frequency. It's those few that I deal with.

2. The concept had to be teachable—in my opinion, at least. I know for a fact that many of the concepts are teachable in such a way that they can be used in scientific and professional pursuits and in everyday life. This is true of many concepts that university courses teach, and I have successfully taught many of those as well as many others in brief laboratory sessions. The remainder of the concepts are similar enough to the ones I know to be teachable that I include them in the book.

3. Most of the concepts form the core of systems of thought. For example, all of the concepts taught in the crucial first-semester statistics course are presented in this book. These concepts are essential for reasoning about a huge variety of problems ranging

from what retirement plan to choose to whether you have enough evidence to decide whether a given job candidate would make a good employee. Taking a course in statistics is not going to help you much in solving those problems, though. Statistics is usually taught in such a way that people can see only that it applies to data of particular, rather limited types. What's needed is what's provided in this book—namely the ability to code events and objects in such a way that rough-and-ready versions of statistical principles can be applied to them. The book also presents the most important concepts of microeconomics and decision theory, the basic principles of the scientific method as they apply to solving everyday problems, the basic concepts of formal logic, the much less familiar principles of dialectical reasoning, and some of the most important concepts developed by philosophers who study how scientists as well as ordinary folks think (or should think).

4. The concepts in the book can be triangulated to understand a given problem from many perspectives. For example, a particularly serious error in everyday life is gross overgeneralization from a small number of observations of a person, object, or event. This error is based on at least four mistakes that compound one another: one psychological, one statistical, one epistemological (epistemology concerns theory of knowledge), and one metaphysical (metaphysics concerns beliefs about the fundamental nature of the world). Once each of these kinds of concepts is well understood, they can all be brought to bear on a given problem, supplementing and enhancing one another.

Every concept in this book is relevant to the way you live your life and conduct your business. We fail to make a friend because we made hasty judgments based on insufficient evidence. We hire people who are not the most capable because we trusted firsthand information too much and more extensive and superior information from other sources too little. We lose money because we don't realize the applicability of statistical concepts such as standard deviation and regression and the relevance of psychological concepts such as the endowment effect, which causes us to want to keep things for no better reason than that we have them, and economic concepts such as sunk costs, which cause us to send good money after bad. We eat foods and take medicines and consume vitamins and other supplements that aren't good for us because we're not sufficiently skilled in evaluating alleged scientific findings about health practices. Society tolerates government and business practices that make our lives worse because they were developed without following effective evaluation procedures and remain untested long after they were introduced—sometimes for decades and at costs in the billions of dollars.

A Sampling of the Things to Come

The first section of the book deals with thinking about the world and ourselves—how we do it, how we flub it, how to fix it, and how we can make far better use than we do of the dark matter of the mind, namely the unconscious.

The second section is about choices—how classical economists think choices are made and how they think they ought to be made, and why modern behavioral economics provides both descriptions of actual choice behavior and prescriptions for it that are better and more useful in some ways than those of classical economics. The section provides suggestions for how to structure your life in order to avoid a wide range of choice pitfalls.

The third section is about how to make categorizations of the world more accurately, how to detect relationships among events, and just as important, how to avoid seeing relationships when they aren't there. Here, we'll examine how to detect errors in reasoning we encounter in the media, at the office, and in bull sessions.

The fourth section is about causality: how to distinguish between cases where one event causes another and cases where events occur close together in time or place but aren't causally related; how to identify the circumstances in which experiments—and only experiments—can make us confident that events are related causally; and how we can learn to be happier and more effective by conducting experiments on ourselves.

The fifth section is about two very different types of reasoning. One of these, logic, is abstract and formal and has always been central to

Western thought. The other, dialectical reasoning, consists of principles for deciding about the truth and practical utility of propositions about the world. This approach to reasoning has always been central to Eastern thought. Versions of the approach have been around in Western thought since the time of Socrates. But only recently have thinkers tried to describe dialectical thought in a systematic way or relate it to the tradition of formal logic.

The sixth section is about what constitutes a good theory about some aspect of the world. How can we be sure that what we believe is actually true? Why is it that simpler explanations are normally more useful than more complicated ones? How can we avoid coming up with slipshod and overly facile theories? How can theories be verified, and why should we be skeptical of any assertion that can't, at least in principle, be falsified?

The sections of the book support one another. Understanding what we can and can't observe about our mental life tells us when to rely on intuition when solving a problem and when to turn to explicit rules about categorization, choice, or assessment of causal explanations. Learning about how to maximize the outcomes of choices depends on what's been learned about the unconscious and how to make it an equal partner with the conscious mind when choosing actions or predicting what will make us happy. Learning about statistical principles tips us off about when we need to reach for our rules for assessing causality. Knowledge of how to assess causality encourages us to trust experiments far more than simple observation of events and shows how important (and how easy) it can be to conduct experiments to tell us what business practices and personal behaviors are most likely to benefit us. Learning about logic and dialectical reasoning provides suggestions for different ways to come up with theories about a given aspect of the world, which in turn can suggest what kinds of methods will be necessary to test those theories.

You won't have a higher IQ when you finish this book, but you'll be smarter.

PART I

THINKING ABOUT THOUGHT

Psychological research has produced three major insights about the way the mind works that will change the way you think about how you think.

The first is the proposition that our understanding of the world is always a matter of *construal*—of inference and interpretation. Our judgments about people and situations, and even our perceptions of the physical world, rely on stored knowledge and hidden mental processes and are never a direct readout of reality. A full appreciation of the degree to which our understanding of the world is based on inferences makes it clear how important it is to improve the tools we use to make those inferences.

Second, the situations we find ourselves in affect our thoughts and determine our behavior far more than we realize. People's *dispositions*, on the other hand—their distinctive traits, attitudes, abilities, and tastes— are much less influential than we assume. So we make mistakes in assessing why it is that people—including ourselves—believe particular things and behave in particular ways. But it's possible to overcome this "fundamental attribution error" to a degree.

Finally, psychologists have increasingly come to recognize the importance of the *unconscious mind*, which registers vastly more environmental information than the conscious mind could possibly notice. Many of the most important influences on our perceptions and behavior are hidden from us. And we are *never* directly aware of the mental processes that produce our perceptions, beliefs, and behavior. Fortunately, and perhaps surprisingly, the unconscious is fully as rational as the conscious mind. It solves many kinds of problems the conscious mind can't deal with effectively. A few simple strategies allow us to harness the unconscious mind's problem-solving capacities.

1. Everything's an Inference

Without a profound simplification the world around us would be an infinite, undefined tangle that would defy our ability to orient ourselves and decide upon our actions . . . We are compelled to reduce the knowable to a schema.   —Primo Levi, *The Drowned and the Saved*

First baseball umpire: "I call 'em as I see 'em."
Second umpire: "I call 'em as they are."
Third umpire: "They ain't nothin' till I call 'em."

When we look at a bird or a chair or a sunset, it feels as if we're simply registering what is in the world. But in fact our perceptions of the physical world rely heavily on tacit knowledge, and mental processes we're unaware of, that help us perceive something or accurately categorize it. We know that perception depends on mental doctoring of the evidence because it's possible to create situations in which the inference processes we apply automatically lead us astray.

Have a look at the two tables below. It's pretty obvious that one of the tables is longer and thinner than the other.

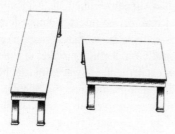

Figure 1. Illusion created by the psychologist Roger Shepard.[1]

Obvious, but wrong. The two tables are of equal length and width.

The illusion is based on the fact that our perceptual machinery decides for us that we're looking at the end of the table on the left and the side of the table on the right. Our brains are wired so that they "lengthen" lines that appear to be pointing away from us. And a good thing, too. We evolved in a three-dimensional world, and if we didn't tamper with the sense impression—what falls on the eye's retina—we would perceive objects that are far away as being smaller than they are. But what the unconscious mind brings to perception misleads us in the two-dimensional world of pictures. As a result of the brain's automatically increasing the size of things that are far away, the table on the left appears longer than it is and the table on the right appears wider than it is. When the objects aren't really receding into the distance, the correction produces an incorrect perception.

Schemas

We aren't too distressed when we discover that lots of unconscious processes allow us to correctly interpret the physical world. We live in a three-dimensional world, and we don't have to worry about the fact that the mind makes mistakes when it's forced to deal with an unnatural, two-dimensional world. It's more unsettling to learn that our understanding of the nonmaterial world, including our beliefs about the characteristics of other people, is also utterly dependent on stored knowledge and hidden reasoning processes.

Meet "Donald," a fictitious person experimenters have presented to participants in many different studies.

Donald spent a great amount of his time in search of what he liked to call excitement. He had already climbed Mt. McKinley, shot the Colorado rapids in a kayak, driven in a demolition derby, and piloted a jet-powered boat—without knowing very much about boats. He had risked injury, and even death, a number of times. Now he was in search of new excitement. He was thinking, perhaps, he would do some skydiving or maybe cross the Atlantic in a sailboat. By the way he acted one could readily guess that Donald was well aware of his ability to do many things well. Other than business engagements, Donald's contacts with people were rather limited. He felt he didn't really need to rely on anyone. Once Donald made up his mind to do something it was as good as done no matter how long it might take or how difficult the going might be. Only rarely did he change his mind even when it might well have been better if he had.[2]

Before reading the paragraph about Donald, participants first took part in a bogus "perception experiment" in which they were shown a number of trait words. Half the participants saw the words "self-confident," "independent," "adventurous," and "persistent" embedded among ten trait words. The other half saw the words "reckless," "conceited," "aloof," and "stubborn." Then the participants moved on to the "next study," in which they read the paragraph about Donald and rated him on a number of traits. The Donald paragraph was intentionally written to be ambiguous as to whether Donald is an attractive, adventurous sort of person or an unappealing, reckless person. The perception experiment reduced the ambiguity and shaped readers' judgments of Donald. Seeing the words "self-confident," "persistent," and so on resulted in a generally favorable opinion of Donald. Those words conjure up a *schema* of an active, exciting, interesting person. Seeing the words "reckless," "stubborn," and so on triggers a schema of an unpleasant person concerned only with his own pleasures and stimulation.

Since the 1920s, psychologists have made much use of the schema concept. The term refers to cognitive frameworks, templates, or rule systems that we apply to the world to make sense of it. The progenitor of the modern concept of schema is the Swiss developmental psychologist Jean Piaget. For example, Piaget described the child's schema for the "conservation of matter"—the rule that the amount of matter is the same regardless of the size and shape of the container that holds it. If you pour water from a tall, narrow container into a short, wide one and ask a young child whether the amount of water is more, less, or the same, the child is likely to say either "more" or "less." An older child will recognize that the amount of water is the same. Piaget also identified more abstract rule systems such as the child's schema for probability.

We have schemas for virtually every kind of thing we encounter. There are schemas for "house," "family," "civil war," "insect," "fast food restaurant" (lots of plastic, bright primary colors, many children, so-so food), and "fancy restaurant" (quiet, elegant decor, expensive, high likelihood the food will be quite good). We depend on schemas for construal of the objects we encounter and the nature of the situation we're in.

Schemas affect our behavior as well as our judgments. The social psychologist John Bargh and his coworkers had college students make grammatical sentences out of a scramble of words, for example, "Red Fred light a ran."[3] For some participants, a number of the words—"Florida," "old," "gray," "wise"—were intended to call up the stereotype of an elderly person. Other participants made sentences from words that didn't play into the stereotype of the elderly. After completing the unscrambling task, the experimenters dismissed the participants. The experimenters measured how rapidly the participants walked away from the lab. Participants who had been exposed to the words suggestive of elderly people walked more slowly toward the elevator than unprimed participants.

If you're going to interact with an old person—the schema for which one version of the sentence-unscrambling task calls up—it's best not to run around and act too animated. (That is, if you have positive attitudes toward the elderly. Students who are not favorably disposed toward the elderly actually walk *faster* after the elderly prime!)[4]

Without our schemas, life would be, in William James's famous words, "a blooming, buzzing confusion." If we lacked schemas for weddings, funerals, or visits to the doctor—with their tacit rules for how to behave in each of these situations—we would constantly be making a mess of things.

This generalization also applies to our *stereotypes*, or schemas about particular types of people. Stereotypes include "introvert," "party animal," "police officer," "Ivy Leaguer," "physician," "cowboy," "priest." Such stereotypes come with rules about the customary way that we behave, or should behave, toward people who are characterized by the stereotypes.

In common parlance, the word "stereotype" is a derogatory term, but we would get into trouble if we treated physicians the same as police officers, or introverts the same as good-time Charlies. There are, however, two problems with stereotypes: they can be mistaken in some or all respects, and they can exert undue influence on our judgments about people.

Psychologists at Princeton had students watch a videotape of a fourth grader they called "Hannah."[5] One version of the video reported that Hannah's parents were professional people. It showed her playing in an obviously upper-middle-class environment. Another version reported that Hannah's parents were working class and showed her playing in a run-down environment.

The next part of the video showed Hannah answering twenty-five academic achievement questions dealing with math, science, and reading. Hannah's performance was ambiguous: she answered some difficult questions well but sometimes seemed distracted and flubbed easy questions. The researchers asked the students how well they thought Hannah would perform in relation to her classmates. The students who saw an upper-middle-class Hannah estimated that she would perform better than average, while those who saw the working-class Hannah assumed she would perform worse than average.

It's sad but true that you're actually more likely to make a correct prediction about Hannah if you know her social class than if you don't. In general, it's the case that upper-middle-class children perform better in school than working-class children. Whenever the direct evidence about a person or object is ambiguous, background knowledge in the form of a schema or stereotype can increase accuracy of judgments to the extent that the stereotype has some genuine basis in reality.

The much sadder fact is that working-class Hannah starts life with two strikes against her. People will expect and demand less of her, and they will perceive her performance as being worse than if she were upper middle class.

A serious problem with our reliance on schemas and stereotypes is that they can get triggered by incidental facts that are irrelevant or misleading. Any stimulus we encounter will trigger *spreading activation* to related mental concepts. The stimulus radiates from the initially activated concept to the concepts that are linked to it in memory. If you hear the word "dog," the concept of "bark," the schema for "collie," and an image of your neighbor's dog "Rex" are simultaneously activated.

We know about spreading activation effects because cognitive psychologists find that encountering a given word or concept makes us quicker to recognize related words and concepts. For example, if you say the word "nurse" to people a minute or so before you ask them to say "true" or "false" to statements such as "hospitals are for sick people," they will say "true" more rapidly than if they hadn't just heard the word "nurse."[6] As we'll see, incidental stimuli influence not only the speed with which we recognize the truth of an assertion but also our actual beliefs and behavior.

But first—about those umpires who started off this chapter. Most of the time we're like the second umpire, thinking that we're seeing the world the way it really is and "calling 'em as they are." That umpire is what philosophers and social psychologists call a "naive realist."[7] He believes that the senses provide us with a direct, unmediated understanding of the world. But in fact, our construal of the nature and meaning of events is massively dependent on stored schemas and the inferential processes they initiate and guide.

We do partially recognize this fact in everyday life and realize that, like the first umpire, we really just "call 'em as we see 'em." At least we see that's true for other people. We tend to think, "I'm seeing the world as it is, and your different view is due to poor eyesight, muddled thinking, or self-interested motives!"

The third umpire thinks, "They ain't nothin' till I call 'em." All "reality" is merely an arbitrary construal of the world. This view has a long history. Right now its advocates tend to call themselves "postmodernists" or "deconstructionists." Many people answering to these labels endorse the idea that the world is a "text" and no reading of it can be held to be any more accurate than any other. This view will be discussed in Chapter 16.

The Way to a Judge's Heart Is Through His Stomach

Spreading activation makes us susceptible to all kinds of unwanted influences on our judgments and behavior. Incidental stimuli that drift into the cognitive stream can affect what we think and what we do, including even stimuli that are completely unrelated to the cognitive task at hand. Words, sights, sounds, feelings, and even smells can influence our understanding of objects and direct our behavior toward them. That can be a good thing or a bad thing, depending.

Which hurricane is likely to kill more people—one named Hazel or one named Horace? Certainly seems it could make no difference. What's in a name, especially one selected at random by a computer? In fact, however, Hazel is likely to kill lots more people.[8] Female-named hurricanes don't seem as dangerous as male-named ones, so people take fewer precautions.

Want to make your employees more creative? Expose them to the Apple logo.[9] And avoid exposing them to the IBM logo.

It's also helpful for creativity to put your employees in a green or blue environment (and avoid red at all costs).[10] Want to get lots of hits on a dating website? In your profile photo, wear a red shirt, or at least put a red border around the picture.[11] Want to get taxpayers to support education bond issues? Lobby to make schools the primary voting location.[12] Want to get the voters to outlaw late-term abortion? Try to make churches the main voting venue.

Want to get people to put a donation for coffee in the honest box? On a shelf above the coffee urn, place a coconut like the one on the left in the picture below. That would be likely to cause people to behave more honestly. An inverted coconut like the one on the right would likely net you nothing. The coconut on the left is reminiscent of a human face (*coco* is Spanish for head) and people subconsciously sense their behavior is being monitored. (Tacitly, of course—people who literally think they're looking at a human face would be in dire need of an optometrist or a psychiatrist, possibly both.)

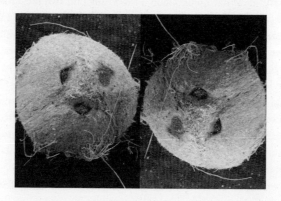

Actually, it's sufficient to just have a picture of three dots in the orientation of the coconut on the left to get more contributions.[13]

Want to persuade someone to believe something by giving them an editorial to read? Make sure the font type is clear and attractive. Messy-looking messages are much less persuasive.[14] But if the person reads the editorial in a seafood store or on a wharf, its argument may be rejected[15] — if the person is from a culture that uses the expression "fishy" to mean "dubious," that is. If not, the fishy smell won't sway the person one way or the other.

Starting up a company to increase IQ in kids? Don't call it something boring like Minnesota Learning Corporation. Try something like FatBrain.com instead. Companies with sexy, interesting names are more attractive to consumers and investors.[16] (But don't actually use FatBrain.com. That's the name of a company that really took off after it changed its drab name to that one.)

Bodily states also find their way into the cognitive stream. Want to be paroled from prison? Try to get a hearing right after lunch. Investigators found that if Israeli judges had just finished a meal, there was a 66 percent chance they would vote for parole.[17] A case that came up just before lunch had precisely zero chance for parole.

Want someone you're just about to meet to find you to be warm and cuddly? Hand them a cup of coffee to hold. And don't by any means make that an iced coffee.[18]

You may recall the scene in the movie *Speed* where, immediately after a harrowing escape from death on a careening bus, two previously unacquainted people (played by Keanu Reeves and Sandra Bullock) kiss each other passionately. It could happen. A man who answers a questionnaire administered by a woman while the two are standing on a swaying suspension bridge high above a river is much more eager to date her than if the interview takes place on terra firma.[19] The study that found this effect is one of literally dozens that show that people can misattribute physiological arousal produced by one event to another, altogether different one.

If you're beginning to suspect that psychologists have a million of these, you wouldn't be far wrong. The most obvious implication of all the evidence about the importance of incidental stimuli is that you want to rig environments so that they include stimuli that will make you or your product or your policy goals attractive. It's obvious when stated that way. Less obvious are two facts: (1) The effect of incidental stimuli can be huge, and (2) you want to know as much as you possibly can about what kinds of stimuli produce what kinds of effects. A book by Adam Alter called *Drunk Tank Pink* is a good compendium of many of the effects we know about to date. (Alter chose the title because of the belief of many prison officials and some researchers that pink walls make inebriated men tossed into a crowded holding cell less prone to violence.)

A less obvious implication of our susceptibility to "incidental" stimuli is the importance of encountering objects—and especially people—in a number of different settings if a judgment about them is to be of any consequence. That way, incidental stimuli associated with given encounters will tend to cancel one another out, resulting in a more accurate impression. Abraham Lincoln once said, "I don't like that man. I must get to know him better." To Lincoln's adage, I'd add: Vary the circumstances of the encounters as much as possible.

Framing

Consider the Trappist monks in two (apocryphal) stories. Monk 1 asked his abbot whether it would be all right to smoke while he prayed. Scandalized, the abbot said, "Of course not; that borders on sacrilege." Monk 2 asked his abbot whether it would be all right to pray while he smoked. "Of course," said the abbot, "God wants to hear from us at any time."

Our construal of objects and events is influenced not just by the schemas that are activated in particular contexts, but by the *framing* of judgments we have to make. The order in which we encounter information of various kinds is one kind of framing. Monk 2 was well aware of the importance of order of input for framing his request.

Framing can also be a matter of choosing between warring labels. And those labels matter not just for how we think about things and how we behave toward them, but also for the performance of products in the marketplace and the outcome of public policy debates.

Your "undocumented worker" is my "illegal alien." Your "freedom fighter" is my "terrorist." Your "inheritance tax" is my "death tax." You are in favor of abortion because you regard it as a matter of exercising "choice." I am opposed because I am "pro-life."

My processed meat, which is 75 percent lean, is more attractive than your product, which has 25 percent fat content.[20] And would you prefer a condom with a 90 percent success rate or one with a 10 percent failure rate? Makes no difference if I pit them against each other as I just did. But students told about the usually successful condom think it's better than do other students told about the sometimes unsuccessful condom.

Framing can affect decisions that are literally a matter of life or death. The psychologist Amos Tversky and his colleagues told physicians about the effects of surgery versus radiation for a particular type of cancer.[21] They told some physicians that, of 100 patients who had the surgery, 90 lived through the immediate postoperative period, 68 were still alive at the end of a year, and 34 were still alive after five years. Eighty-two percent of physicians given this information recommended surgery. Another group of physicians were given the "same" information but in a different form. The investigators told them that 10 of 100 patients died during surgery or immediately after, 32 had died by the end of the year, and 66 had died by the end of five years. Only 56 percent of physicians given this version of the survival information recommended surgery. Framing can matter. A lot.

A Cure for Jaundice

We often arrive at judgments or solve problems by use of *heuristics*—rules of thumb that suggest a solution to a problem. Dozens of heuristics have been identified by psychologists. The effort heuristic encourages us to assume that projects that took a long time or cost a lot of money are more valuable than projects that didn't require so much effort or time. And in fact that heuristic is going to be helpful more often than not. A price heuristic encourages us—mostly correctly—to assume that more expensive things are superior to things of the same general kind that are less expensive. A scarcity heuristic prompts us to assume that rarer things are more expensive than less rare things of the same kind. A familiarity heuristic causes Americans to estimate that Marseille has a bigger population than Nice and Nice has a bigger population than Toulouse. Such heuristics are helpful guides for judgment—they'll often give us the right answer and normally beat a stab in the dark, often by a long shot. Marseille does indeed have a bigger population than Nice. But Toulouse has a bigger population than Nice.

Several important heuristics were identified by the Israeli cognitive psychologists Amos Tversky and Daniel Kahneman.

The most important of their heuristics is the *representativeness heuristic*.[22] This rule of thumb leans heavily on judgments of similarity. Events are judged as more likely if they're similar to the prototype of the event than if they're less similar. The heuristic is undoubtedly helpful more often than not. Homicide is a more representative cause of death than is asthma or suicide, so homicides seem more likely causes than asthma or suicide. Homicide is indeed a more likely cause of death than asthma, but there are twice as many suicide deaths in the United States in a given year than homicide deaths.

Is she a Republican? In the absence of other knowledge, using the representativeness heuristic is about the best we can do. She is more similar to—representative of—my stereotype of Republicans than my stereotype of Democrats.

A problem with that kind of use of the representativeness heuristic is that we often have information that should cause us to assign less weight to the similarity judgment. If we meet the woman at a chamber of commerce lunch, we should take that into account and shift our guess in the Republican direction. If we meet her at a breakfast organized by Unitarians, we should shift our guess in the Democrat direction.

A particularly unnerving example of how the representativeness heuristic can produce errors concerns one "Linda." "Linda is thirty-one years old, single, outspoken, and very bright. She majored in philosophy. As a student, she was deeply concerned with issues of discrimination and social justice and also participated in antinuclear demonstrations." After reading this little description, people were asked to rank eight possible futures for Linda.[23] Two of these were "bank teller" and "bank teller and active in the feminist movement." Most people said that Linda was more likely to be a bank teller active in the feminist movement than just a bank teller. "Feminist bank teller" is more similar to the description of Linda than "bank teller" is. But of course this is a logical error. The *conjunction* of two events can't be more likely than just one event by itself. Bank tellers include feminists, Republicans, and vegetarians. But the description of Linda is more nearly representative of a feminist bank teller than of a bank teller, so the conjunction error gets made.

Examine the four rows of numbers below. Two were produced by a random number generator and two were generated by me. Pick out the two rows that seem to you to be most likely to have been produced by a random number generator. I'll tell you in just a bit which two they are.

```
1 1 0 0 0 1 1 1 1 1 1 0 0 1 0 0 1 0 0 1
1 1 0 0 0 0 0 1 0 1 0 1 0 1 0 1 0 0 0 0 0
1 0 1 0 1 1 1 1 0 1 0 1 0 0 0 1 1 1 0 1 0
0 0 1 1 0 0 0 1 1 0 1 0 0 0 0 1 1 1 0 1 1
```

Representativeness judgments can influence all kinds of estimates about probability. Kahneman and Tversky gave the following problem to undergraduates who hadn't taken any statistics courses.[24]

A certain town is served by two hospitals. In the larger hospital about forty-five babies are born each day, and in the smaller hospital about fifteen babies are born each day. As you know, about 50 percent of all babies are boys. The exact percentage of baby boys, however, varies from day to day. Sometimes it may be higher than 50 percent, sometimes lower.

For a period of one year, each hospital recorded the days on which more than 60 percent of the babies born were boys. Which hospital do you think recorded more such days?

Most of the students thought that the percent of babies that were boys would be the same in the two hospitals. As many thought it would be the larger hospital that would have the higher percentage as thought it would be the smaller hospital.

In fact, it's vastly more likely that percentages of sixty-plus for boys would occur in the small hospital. Sixty percent is equally representative (or, rather, nonrepresentative) of the population value whether the hospital is small or large. But deviant values are far more likely when there are few cases than when there are many.

If you doubt this conclusion, try this. There are two hospitals, one with five births per day and one with fifty. Which hospital do you think would be expected to have 60 percent or more boy babies on a given day? Still recalcitrant? How about five babies versus five thousand?

The representativeness heuristic can affect judgments of the probability of a limitless number of events. My grandfather was once a well-to-do farmer in Oklahoma. One year his crops were ruined by hail. He had no insurance, but he didn't bother to get any for the coming year because it was so unlikely the same thing would happen two years in a row. That's an unrepresentative pattern for hail. Hail is a rare event and so *any* particular sequence of hail is unlikely. Unfortunately, hail doesn't remember whether it happened last year in northwest Tulsa or in southeast Norman. My grandfather did get hailed out the next year. He didn't bother to get insurance for the next year because it was really inconceivable that hail would strike the same place three years in a row. But that in fact did happen. My grandfather was bankrupted by his reliance on the representativeness heuristic to judge probabilities. As a consequence, I'm a psychologist rather than a wheat baron.

Back to those rows of numbers I asked you about earlier. It's the top two rows that are genuinely random. They were two of the first three sequences I pulled from a random number generator. Honest. I did not cherry-pick beyond throwing out the one sequence. The last two rows I made up because they're more representative of a random sequence than random sequences are. The problem is that our conception of the randomness prototype is off kilter. Random sequences have too many more long runs (00000) and too many more regularities (01010101) than they "should." Bear this in mind when you see a basketball player score points five times in a row. There's no reason to keep passing the ball to him any more than to some other player. The player with the "hot hand" is no more likely to make the shot than another player with a comparable record for the season.[25] (The more familiar you are with basketball, the less likely you are to believe this. The more familiar you are with statistics and probability theory, the more likely you are to believe it.)

The basketball error is characteristic of a huge range of mistaken inferences. Simply put, we see patterns in the world where there are none because we don't understand just how un-random-looking random sequences can be. We suspect the dice roller of cheating because he gets three 7s in a row. In fact, three 7s are much more likely than 3, 7, 4, which would not likely arouse suspicion. We hail a friend as a stock guru because all four of the stocks he bought last year did better than the market as a whole. But four hits is no less likely to happen by chance than two hits and two misses or three hits and one miss. So it's premature to hand over your portfolio to your friend. The representativeness heuristic sometimes influences judgments about causality. I don't know whether Lee Harvey Oswald acted alone in the assassination of John F. Kennedy or whether there was a conspiracy involving other people. I have no doubt, though, that part of the reason so many people have been convinced that there was a conspiracy is that they find it implausible that an event of such magnitude could have been effected by a single, quite unprepossessing individual acting alone.

Some of the most important judgments about causality that we make concern the similarity of a disease and treatment for the disease. The Azande people of Central Africa formerly believed that burnt skull of the red bush monkey was an effective treatment for epilepsy. The jerky, frenetic movements of the bush monkey resemble the convulsive movements of epileptics.

The Azande belief about proper treatment for epilepsy would have seemed sensible to Western physicians until rather recently. Eighteenth-century doctors believed in a concept called the "doctrine of signatures." This was the belief that diseases could be cured by finding a natural substance that resembles the disease in some respect. Turmeric, which is yellow, would be effective in treating jaundice, in which the skin turns yellow. The lungs of the fox, which is known for strong powers of respiration, were considered a remedy for asthma.

The belief in the doctrine of signatures was derived from a theological principle: God wishes to help us find the cures for diseases and gives us helpful hints in the form of color, shape, and movement. He knows we expect the treatment to be representative of the illness. This now sounds dubious to most of us, but in fact the representativeness heuristic continues to underlie alternative medicine practices such as homeopathy and Chinese traditional medicine—both of which are increasing in popularity in the West.

Representativeness is often the basis for predictions when other information would actually be more helpful. About twenty years out from graduate school a friend and I were talking about how successful our peers had been as scientists. We were surprised to find how wrong we were about many of them. Students we thought were sure to do great things often turned out to have done little in the way of good science; students we thought were no great shakes turned out to have done lots of excellent work. In trying to figure out why we could have been so wrong, we began to realize that we had relied on the representativeness heuristic. Our predictions were based in good part on how closely our classmates matched our stereotype of an excellent psychologist—brilliant, well read, insightful about people, fluent. Next we tried to see whether there was any way we could have made better predictions. It quickly became obvious: the students who had done good work in graduate school did good work in their later career; those who hadn't fizzled.

The lesson here is one of the most powerful in all psychology. The best predictor of future behavior is past behavior. You're rarely going to do better than that. Honesty in the future is best predicted by honesty in the past, not by whether a person looks you steadily in the eye or claims a recent religious conversion. Competence as an editor is best predicted by prior performance as an editor, or at least by competence as a writer, and not by how verbally clever a person seems or how large the person's vocabulary is.

Another important heuristic Tversky and Kahneman identified is the *availability heuristic.* This is a rule of thumb we use to judge the frequency or plausibility of a given type of event. The more easily examples of the event come to mind, the more frequent or plausible they seem. It's a perfectly helpful rule most of the time. It's easier to come up with the names of great Russian novelists than great Swedish novelists, and there are indeed more of the former than the latter. But are there more tornadoes in Kansas or in Nebraska? Pretty tempting to say Kansas, isn't it? Never mind that the Kansas tornado you're thinking about never happened.

Are there more words with the letter *r* in the first position or the third position? Most people say it's the first position. It's easier to come up with words beginning with *r* than words having an *r* in the third position—because we "file" words in our minds by their initial letters and so they're more available as we rummage through memory. But in fact there are more words with *r* in the third position.

One problem with using the availability heuristic for judgments of frequency or plausibility is that availability is tangled up with salience.

Deaths by earthquake are easier to recall than deaths by asthma, so people overestimate the frequency of earthquake deaths in their country (by a lot) and underestimate the frequency of asthma deaths (hugely).

Heuristics, including the representativeness heuristic and the availability heuristic, operate quite automatically and often unconsciously. This means it's going to be hard to know just how influential they can be. But knowing about them allows us to reflect on the possibility that we've been led astray by them in a particular instance.

Summing Up

It's possible to make fewer errors in judgment by following a few simple suggestions implicit in this chapter.

Remember that all perceptions, judgments, and beliefs are inferences and not direct readouts of reality. This recognition should prompt an appropriate humility about just how certain we should be about our judgments, as well as a recognition that the views of other people that differ from our own may have more validity than our intuitions tell us they do.

Be aware that our schemas affect our construals. Schemas and stereotypes guide our understanding of the world, but they can lead to pitfalls that can be avoided by recognizing the possibility that we may be relying too heavily on them. We can try to recognize our own stereotype-driven judgments as well as recognize those of others.

Remember that incidental, irrelevant perceptions and cognitions can affect our judgment and behavior. Even when we don't know what those factors might be, we need to be aware that much more is influencing our thinking and behavior than we can be aware of. An important implication is that it will increase accuracy to try to encounter objects and people in as many different circumstances as possible if a judgment about them is important.

Be alert to the possible role of heuristics in producing judgments. Remember that the similarity of objects and events to one another can be a misleading basis for judgments. Remember that causes need not resemble effects in any way. And remember that assessments of the likelihood or frequency of events can be influenced simply by the readiness with which they come to mind.

Many of the concepts and principles you're going to read about in this book are helpful in avoiding the kinds of inferential errors discussed in this chapter. These new concepts and principles will supplement, and sometimes actually replace, those you normally use.

2. The Power of the Situation

The previous chapter showed that we're frequently ignorant about the influence of irrelevant, incidental, and scarcely noticed stimuli in producing our judgments and behavior. Unfortunately, we're also frequently blind to the role played by factors that are not incidental or evanescent at all but rather are the prime movers affecting our judgments and behavior. In particular, we often underestimate—or fail to notice at all—some of the most important situational influences that markedly affect beliefs and behavior.

A direct consequence of this "context blindness" is that we tend to exaggerate the influence of personal, "dispositional" factors—preferences, personality traits, abilities, plans, and motives—on behavior in a given situation.

Both the slighting of the situation and the exaggeration of internal factors occur even when we're trying to analyze the reasons for our own judgments and the causes of our own behavior. But the problem is much greater when it's the causes of other people's behavior we're trying to figure out. I have to attend to many aspects of the context and situation if I'm going to be able to form a judgment or carry out some behavior. But the situation confronting another person may be difficult or impossible for me to see. So I'm particularly likely to underestimate the importance of the situation for another's behavior and to overestimate internal factors.

The failure to recognize the importance of contexts and situations and the consequent overestimation of the role of personal dispositions is, I believe, the most pervasive and consequential inferential mistake we make. The social psychologist Lee Ross has labeled this the *fundamental attribution error*.

As it happens, there are big cultural differences in propensity to make this error. This fact offers the hope that people in more susceptible cultures may be able to overcome the error to some degree.

The Fundamental Attribution Error

Bill Gates is the richest person in the world. At the ripe old age of nineteen, Gates dropped out of Harvard to start Microsoft, and in a few short years he made it the most profitable corporation in the world. It's tempting to think that he must be one of the smartest people who ever lived.

Gates is undoubtedly extraordinarily bright. But what few people know about him is that his precollege life was blessed, computationally speaking. He was bored at his Seattle public school in eighth grade in 1968, so his parents switched him to a private school that happened to have a terminal linked to a mainframe computer. Gates became one of a small number of people anywhere who had substantial time to explore a high-powered computer. His luck continued for the next six years. He was allowed to have free programming time in exchange for testing the software of a local company; he regularly sneaked out of his house at three in the morning to go to the University of Washington computer center to take advantage of machine time made available to the public at that hour. There was not likely another teenager in the world who had the kind of access to computers that Gates had.

Behind many a successful person lies a string of lucky breaks that we have no inkling about. The economist Smith has twice as many publications in refereed journals as the economist Jones. We're naturally going to assume that Smith is more talented and hardworking than Jones. But as it happens, economists who get their PhDs in a "fat year," when there are many university jobs available, do much better in the academic job market and have more successful careers than economists who get their PhDs in a "lean year." The difference in success between Smith and Jones may have more to do with dumb luck than with smarts, but we're not going to see this.

The careers of many college students who got their degrees during the Great Recession are going to be forever stunted. Unemployment is bad not just because it's demoralizing not to have a job, but because the repercussions may never cease. Parents are going to wonder where they went wrong with struggling Jane, who graduated from college in 2009, and what they did that was so different from how they brought up successful Joan, who graduated in 2004.

Important influences can be hidden, but even when powerful situational determinants of behavior are staring us in the face, we can be oblivious to their impact.

In a classic experiment from the 1960s, the social psychologists Edward Jones and Victor Harris showed people one of two essays about Cuba's political system allegedly written by a college student in response to a requirement by a professor.[1] One essay was favorable toward Cuba and the other was unfavorable. The experimenters informed the participants who read the essay favorable to Cuba that it had been written as an assignment: an instructor in a political science course (or, in another experiment, a debate coach) required the student to write a pro-Cuba essay. The experimenters told other participants that the student who wrote the unfavorable essay had been required to write an anti-Cuba essay. I think we can agree that the participants had learned nothing about the students' actual attitudes toward Cuba. Yet the participants rated the first student as being substantially more favorable to Cuba than the second student.

In everyday life we ignore equally powerful influences on people's behavior. A professor friend of mine regularly teaches two different courses to undergraduates at Stanford. One is a statistics course and the other is a community outreach course. The students who take his statistics course rate him at the end of the term as being rigid, humorless, and rather cold. The students who take the community outreach course rate him as flexible, funny, and quite warm.

Whether you're heroic or heartless may depend on a contextual factor whose impact is far greater than we would tend to assume. The social psychologists John Darley and Bibb Latané conducted a series of experiments studying what has come to be known as "bystander intervention."[2] They contrived a number of situations that seemed like emergencies— an epileptic seizure, a bookcase falling on a person in an adjacent room, someone who fainted on the subway. The likelihood of a person offering help to the "victim" was hugely dependent on the presence of others. If people thought they were the only witness, they usually attempted to help.

If there was another "witness" (actually a confederate of the experimenter), they were much less likely to help. If there were many "witnesses," people were quite unlikely to offer help.

In Darley and Latané's "seizure" experiment, in which people thought they were communicating over an intercom, 86 percent of people rushed to help the "victim" when they thought they were the only person who knew about the incident. If they thought there were two bystanders, 62 percent of people offered help. When four people presumably heard the cries for help, only 31 percent volunteered their services.

To drive home the point that kindness and caring can be less important than situational factors, Darley and his colleague Daniel Batson conducted a study with theological students—people we might assume would be particularly likely to help someone in need.[3] The researchers sent a number of Princeton theological students to a building across campus to deliver a sermon on the Good Samaritan (!), telling them the route to follow. Some of the students were told they had plenty of time to get to the building and others were told they were already late. On their way to deliver the sermon, each of the seminarians passed a man who was sitting in a doorway, head down, groaning and coughing and in obvious need of help. Almost two-thirds of the seminarians offered help to the man if they were in no rush. Only 10 percent offered help if they were late.

Of course, if you knew only that a particular seminarian helped and another one didn't, you would have a much more favorable impression of the one who offered assistance than of the one who didn't. A circumstance like being in a rush wouldn't likely occur to you as a factor influencing the seminarian who failed to be a Good Samaritan. And in fact, when you describe the experimental setup to people, they don't think that the situation—being late versus not—would have any effect at all on whether the seminarian would help or ignore the person in distress.[4] Given this belief, they can only perceive failure to help as being due to poor character, something internal to the person.

Hidden situational factors can also influence how smart a person seems to be. The social psychologist Lee Ross and his colleagues invited students to participate in a study with a TV quiz show format. One student, selected at random, was to ask the questions and the other student was to answer them. The questioner's role was to generate ten "challenging but not impossible questions," and the "contestant" was supposed to provide answers out loud. Questioners took advantage of their role to display esoteric knowledge in the questions they posed. "What is the sweet-smelling waxy stuff that comes from whales and is used as a base for perfume?" (Ambergris—in case you haven't recently read *Moby-Dick*.) Contestants managed to answer only a fraction of the questions.

At the end of the session, both of the participants, as well as the observers, were required to rate both the questioner's and the contestant's general knowledge. You might think that it would have been clear to subjects and observers alike that the questioner's role gave him a big advantage. The role guaranteed that he would reveal no area of ignorance, whereas the contestant's role offered no opportunity for such selective, self-serving displays. But the role advantage of the questioner was not sufficiently obvious, either to the contestants or to the observers, to prevent them from judging the questioners to be unusually knowledgeable. Both the contestants and the observers rated the questioner as far more knowledgeable than either the contestant or the "average" student in the university.

The quiz study has profound relevance to everyday life. The organizational psychologist Ronald Humphrey set up a laboratory microcosm of a business office.[5] He told participants he was interested in "how people work together in an office setting." A showily random procedure selected some of the participants to be "managers" and to assume supervisory responsibilities. Some were selected to be mere "clerks" who followed orders. Humphrey gave the managers time to study manuals describing their tasks. While they were studying them, the experimenter showed the clerks the mailboxes, filing system, and so on. The newly constructed office team then went about their business for two hours. The clerks were assigned to work on a variety of low-skilled, repetitive jobs and had little autonomy. The managers, as in a real office, performed reasonably high-skill-level tasks and directed the clerks' activities.

At the end of the work period, managers and clerks rated themselves and each other on a variety of role-related traits. These included leadership, intelligence, motivation for hard work, assertiveness, and supportiveness. For all these traits, managers rated their fellow managers more highly than they rated their clerks. For all but hardworkingness, clerks rated their managers more highly than they rated their fellow clerks.

People can find it hard to penetrate beyond appearances and recognize the extent to which social roles affect behavior, even when the random basis of role assignment and the prerogatives of particular roles are made abundantly clear. And, of course, in everyday life it's often less clear why people occupy the roles they do, so it can be very difficult to separate role demands and advantages from the intrinsic attributes of the occupant of the role.

Only after I read about these experiments did I understand why I was typically so impressed with the astute questions my colleagues asked in the final oral examinations of PhD candidates—and usually somewhat disappointed by my students' less than trenchant answers!

The fundamental attribution error gets us in trouble constantly. We trust people we ought not to, we avoid people who really are perfectly nice, we hire people who are not all that competent—all because we fail to recognize situational forces that may be operating on the person's behavior. We consequently assume that future behavior will reflect the dispositions we infer from present behavior. (So that you won't think this generalization fails to square with the assertion that past behavior is the best guide to future behavior, note that it's past behavior over the long run, observed in many diverse situations, that is the excellent predictor, not behavior observed in only a few situations, especially a few situations all of the same type.)

Why Do Some Kids Sell Drugs and Other Kids Go to College?

> You are the average of the five people you spend the most time with.
> —Jim Rohn, American entrepreneur and motivational speaker

When my son was fifteen years old, I happened to be looking out my office window when I saw him walking across a parking lot with another boy. They were both smoking cigarettes, which was something my wife and I assumed that my son didn't do and wouldn't do. That evening I said to my son, "I was disappointed to see you smoking a cigarette today." "Yes, I was smoking," he said defiantly. "But it wasn't because of peer pressure."

Yes it was. Or at any rate, he was smoking because a lot of his peers were smoking. We do things all the time because other people are doing them. They model behavior for us and often encourage us, openly or tacitly, to follow their example. They can be successful beyond our imagining.

Social influence is perhaps the most researched topic in all social psychology. We can be blind to it not only when we're observing other people's behavior but when we're trying to explain to ourselves the causes of our own behavior.

The first social psychology experiment was conducted by Norman Triplett in 1898.[6] He found that cyclists had much better times when they competed against another cyclist than when they merely competed against the clock. The general point has been made in scores of subsequent experiments. People perform more energetically not just when they're in competition with others but even when other people are merely observing. The *social facilitation effect* on performance has even been found for dogs, possums, armadillos, frogs, and fish.

(You're probably wondering whether the effect is found for cockroaches as well. It is indeed! The social psychologist Robert Zajonc had cockroaches run for cover when he turned on a light. A cockroach ran faster if there was another cockroach next to it. The cockroaches ran faster even if other cockroaches were mere observers, watching from specially constructed cockroach bleachers.)

Many years ago, I bought a Saab automobile, and shortly thereafter started noticing that several of my colleagues were driving Saabs. Later, my wife and I started playing tennis and were surprised that lots of my friends and acquaintances had taken up tennis as well. After a few years, we drifted away from tennis. I began to notice that the tennis courts I had frequented, instead of being lined with people waiting their turn to play, were standing mostly empty. We took up cross-country skiing—at about the same time that several of our friends did. That, too, we eventually lost interest in. I subsequently noticed that most of my skiing friends had also more or less dropped skiing. And I won't even bother to tell you about serving after-dinner drinks, minivans, going to obscure art films . . .

I was quite unaware of what I can now see as the influence of our friends and neighbors on the behavior of my wife and me. But at the time I would have said the favorable rating *Consumer Reports* gave to Saab was the main reason I bought one. My wife and I wanted to have regular exercise, and there was a tennis court across from our house, so that seemed the natural exercise to take up. There were always things to attribute our behavior to other than the influence of our acquaintances.

We should choose our acquaintances carefully because we're going to be highly influenced by them. This is especially true for young people: the younger you are, the more influenced you are by peers' attitudes and behaviors.[7] One of a parent's most important and challenging roles is to make sure their children's acquaintances are likely to be good influences.

The economists Michael Kremer and Dan Levy examined the grade point averages of students whose freshman roommate had been assigned to them at random.[8] The investigators found out how much alcohol each student had tended to consume in high school. Students who had been assigned a roommate who came to college with a history of substantial drinking got grades a quarter point lower than students assigned a teetotaler. That can easily mean a GPA of B plus versus A minus or C plus versus B minus. If the student himself had been a drinker prior to going to college, that student got grades a full point lower if his roommate had been a drinker than if he had not! That can mean a good medical school for him versus no medical school. (I use the word "him" deliberately; there was no effect on females of having a drinking roommate.)

It seems highly improbable that the unsuspecting student would have recognized that his roommate's drinking was the main cause of his disappointing scholastic achievement. Indeed, the investigators themselves don't know exactly why a roommate's behavior should be so important, though it seems likely that the drinking roommate just made drinking seem like a natural pastime. And of course the more you drink, the less you study, and the less effective you are when you do study.

You can reduce college students' drinking, incidentally, by simply telling them how much drinking goes on at their school.[9] This tends to be substantially less than students think, and they move their drinking in line with that of their peers.

> I understand why [President Obama] wants you to go to college. He wants
> to remake you in his image.
>   —Senator Rick Santorum, during his 2012 presidential campaign

So was Senator Santorum right about what college does to people? Does it really push them toward President Obama's political camp?

Indeed it does. The economist Amy Liu and her colleagues conducted a study of students at 148 different colleges and universities—large and small, public and private, religious and secular.[10] They found that the number of students at the end of college who described themselves as liberal or far left in their politics increased by 32 percent over the number who described themselves as such when they were incoming freshmen. The number who described themselves as conservative or far right decreased by 28 percent. Students moved left on questions of marijuana legalization, same-sex marriage, abortion, abolition of the death penalty, and increasing taxes for the wealthy. If fewer people went to college, Republicans would win more elections.

It's likely that you moved left in college, too. If so, does it seem to you that the liberalism of your professors was responsible? A desire to adopt the views of prestigious older students? I would bet not. It seemed to me that my own leftward drift in college was the consequence not of spongelike absorption of professors' views or slavish imitation of my fellow students, but rather the result of coming independently to a better understanding of the nature of society and the kinds of things that improve it.

But of course, my leftward drift was indeed the result in good part of social influence from students and professors. And those professors were influencing not just their students but each other as well. A conservative student group has claimed that publicly available numbers from the Federal Election Commission showed that, in 2012, 96 percent of Ivy League professors' political donations went to President Obama. They reported that precisely one professor at Brown University gave money to Mitt Romney. (And it may have been sheer cussedness rather than political convictions that prompted the donation!)

Those political contribution trends may be exaggerated, but as a social psychologist and former Ivy League professor, I can assure you that those professors (a) are indeed overwhelmingly liberal and (b) don't recognize the conformity pressures influencing their own opinions. Left to themselves, you wouldn't find 96 percent of Ivy League professors reporting that they think daily tooth brushing is a good idea.

Other institutions are also hothouses of liberalism. A Republican Party operative trying to recruit techies from Google discovered that people were much more likely to be out as gay than out as Republican.

Needless to say, some communities are undoubtedly equally success-ful at fostering and enforcing conservatism. My candidates would include Bob Jones University and the Dallas Chamber of Commerce.

And of course, the whole country isn't moving drastically to the left with each successive generation. Students from those liberal colleges are going to reenter a world of people with a wide range of views—which will now begin to influence them in a more rightward direction on average.

It's not just attitudes and ideology that are influenced by other people. Engage in a conversation with someone in which you deliberately change your bodily position from time to time. Fold your arms for a couple of minutes. Shift most of your weight to one side. Put one hand in a pocket. Watch what your conversation partner does after each change and try not to giggle. "Ideomotor mimicry" is something we engage in quite un-consciously. When people don't do it, the encounter can become awk-ward and unsatisfying.[11] But the participants won't know what it is that went wrong. Instead: "She's kind of a cold fish"; "we don't share much in common."

Awareness of Social Influence

The social psychologists George Goethals and Richard Reckman con-ducted the granddaddy of all studies showing the power of social influ-ence, together with the blissful absence of any awareness of it.[12] They asked white high school students their opinions about a large number of social issues, including one that was very salient and very controversial in their community at the time, namely busing for the purpose of racial integration.[13] A couple of weeks later the investigators called the partici-pants and asked them to participate in a discussion of the busing issue. Each group comprised four participants. Three of the group's participants in a given group were like-minded. Either those members had all indi-cated that they were probusing or they had all indicated that they were antibusing. The fourth person assigned to each group was a ringer em-ployed by the experimenters, armed with a number of persuasive argu-ments against the other group members' opinion. After the discussion, the participants filled out another questionnaire with a different format. One question asked their opinion on the busing issue.

The original antibusing students shifted their position substantially

in a probusing direction. Most of the probusing students were actually
converted to an antibusing position. The investigators asked the participants to recall, as best they could, what their original opinions on the
busing question had been. But first, the investigators reminded the participants that they were in possession of the original opinion scale and
would check the accuracy of the participants' recall. Participants who had
not been asked to participate in a discussion were able to recall their
original opinions with high accuracy. But among the members of the
discussion groups, the original antibusing participants "recalled" their
opinions as having been much more probusing than they actually were.
The original probusing participants actually recalled their original opinions as having been, on average, antibusing!

As well as showing massive social influence and near-total failure to
recognize it, the Goethals and Reckman study also makes the disconcerting and important point that our attitudes about many things, including some very important ones, are not pulled out of a mental file
drawer but rather are constructed on the fly. Just as disconcerting, our
beliefs about our past opinions are also often fabricated. I have a friend
who told me in 2007 he would vote for any of the Republican candidates
over the faddish and untried Obama. When I reminded him of this just
before he enthusiastically voted for Obama in 2008, he was angry that I
could have concocted such a story. I'm frequently told that a current
strongly held opinion of mine conflicts with one I expressed in the past.
When that happens, I can find it impossible to reconstruct the person—
namely me—who could have expressed that opinion.

Actor-Observer Differences in Assessing
the Causes of Behavior

A few years ago, a graduate student who was working with me told me
something about himself that I would never have guessed. He had done
prison time for murder. He hadn't pulled the trigger, but he had been
present when an acquaintance committed the murder, and he was convicted of being an accessory to the crime.

My student told me a remarkable thing about the murderers he met
in prison. To a man, they attributed their homicides to the situation they
had been in. "So I tell the guy behind the counter to give me everything

in the register and instead he reaches under the counter. Of course I had to plug him. I felt bad about it."

There are obvious self-serving motives behind such attributions. But it's important to know that people generally think that their *own* behavior is largely a matter of responding sensibly to the situation they happen to be in—whether that behavior is admirable or abominable. We're much less likely to recognize the situational factors *other people* are responding to, and we're consequently much more likely to commit the fundamental attribution error when judging them—seeing dispositional factors as the main or sole explanation for the behavior.

If you ask a young man why he dates the girl he does, he's likely to say something like, "She's a very warm person." If you ask that same young man why an acquaintance dates the girl he does, he's likely to say, "Because he needs to have a nonthreatening girlfriend."[14]

When you ask people to say whether their behavior, or their best friend's, usually reflects personality traits or whether their behavior depends primarily on the situation, they'll tell you that their friend's behavior is more likely to be consistent across different situations than their own is.[15]

The main reason for differences in the attributions actors and observers make is that the context is always salient for the actor. I need to know what the important aspects of my situation are in order to behave adaptively (though of course I'm going to miss or ignore many important things). But *you* don't have to pay such close attention to the situation that I confront. Instead, what's most salient to you is my behavior. And it's an easy jump from a characterization of my behavior (nice or nasty) to a characterization of my personality (kindly or cruel). You often can't see—or may ignore—important aspects of my situation. So there are few constraints on your inclination to attribute my behavior to my personality.

Culture, Context, and the Fundamental Attribution Error

People who grew up in Western culture tend to have considerable scope and autonomy in their lives. They can often pursue their interests while paying little attention to other people's concerns. People in many other cultures lead more constrained lives. The freedom of the West begins with the remarkable sense of personal agency of the ancient Greeks. In contrast, the equally ancient and advanced civilization of China placed much more emphasis on harmony with others than on freedom of individual action. In China, effective action always required smooth interaction with others—both superiors and peers. The differences between West and East in degree of independence versus interdependence remain today.

In a book called *The Geography of Thought*, I proposed that these different social orientations were economic in origin.[16] Greek livelihoods were based on relatively solitary occupations such as trading, fishing, and animal husbandry, and on agricultural practices such as kitchen gardens and olive tree plantations. Chinese livelihoods were based on agricultural practices, especially rice cultivation, requiring much more cooperation. Autocracy (often benevolent, sometimes not) was perhaps an efficient way of running a society where every man for himself was not an option.

So it was necessary for Chinese to pay attention to social context in a way that it wasn't for Greeks. The differences in attention have been demonstrated in a dozen different ways by experiments conducted with the Western inheritors of Greek independence and the Eastern inheritors of Confucian Chinese traditions. One of my favorite experiments, conducted by the social psychologist Takahiko Masuda, asks Japanese and American college students to rate the expression of the central figure in the cartoon below.[17]

Japanese students rate the central figure as less happy when he's surrounded by sad figures (or angry figures) than when he's surrounded by happier figures. The Americans were much less affected by the emotion of the surrounding figures. (The experiment was also carried out with sad or angry figures in the center and with happy, sad, or angry faces in the background, with similar results.)

The attention to context carries through to physical context. To see how deep this difference in attention to context goes, take a look at the scene below, which is a still from a twenty-second color video of an underwater scene. Masuda and I have shown such videos to scores of people and then asked them to tell us what they saw.[18]

Americans are likely to start off by saying, "I saw three big fish swimming off to the left; they had pink fins and white bellies and vertical stripes on their backs." Japanese are much more likely to say, "I saw what looked like a stream, the water was green, there were rocks and shells on the bottom, there were three big fish swimming off to the left." Only after the context was established did the Japanese zoom in on what are the most salient objects for Americans. Altogether, the Japanese reported seeing 60 percent more background objects than did the Americans. That's what you'd expect, given that East Asians pay more attention to context than do Westerners.

The differential attention to context results in Easterners' having a preference for situational explanations for behavior that Westerners are more likely to explain in dispositional terms. A study by Korean social psychologists found that if you tell someone that a particular person behaved as did most people in the person's situation, Koreans infer, quite reasonably, that something about the situation was the primary factor motivating the person's behavior.[19] But Americans will explain the person's behavior in terms of the person's dispositions—ignoring the fact that others behaved in the same way in the situation.

Easterners are susceptible to the fundamental attribution error, just not as susceptible as Westerners. For example, in a study similar to the one by Jones and Harris demonstrating that people tend to assume an essay writer holds the opinion required by the assignment, Incheol Choi and his coworkers showed that Korean participants made the same mistake as Americans.[20] But when participants were put through the same kind of coercive situation as those whose essays they were about to read, the Koreans got the point and didn't assume that the writer's real attitudes corresponded to their essay position. Americans, however, learned nothing from having the situation made so obvious and assumed they had learned something about the essay writer's opinion.

Easterners tend to have a *holistic* perspective on the world.[21] They see objects (including people) in their contexts, they're inclined to attribute behavior to situational factors, and they attend closely to relationships between people and between objects. Westerners have a more *analytic* perspective. They attend to the object, notice its attributes, categorize the object on the basis of those attributes, and think about the object in terms of the rules that they assume apply to objects of that particular category.

Both perspectives have their place. I have no doubt that the analytic perspective has played a role in Western dominance in science. Science is at base a matter of categorization and discovering the rules that apply to the categories. And in fact, the Greeks invented science at a time when Chinese civilization, though making great progress in mathematics and many other fields, had no real tradition of science in the modern sense.

But the holistic perspective saves Easterners from some serious errors in understanding why other people behave as they do. Moreover, the reluctance to make dispositional attributions contributes to Eastern belief in the capacity of people to change. As we'll see in Chapter 14 on dialectical reasoning, the assumption of malleability of human behavior helps Asians to be correct about important questions that the Western perspective gets wrong.

Summing Up

One of the main lessons of these first two chapters is that there is vastly more going on in our heads than we realize. The implications of this research for everyday life are profound.

Pay more attention to context. This will improve the odds that you'll correctly identify situational factors that are influencing your behavior and that of others. In particular, attention to context increases the likelihood that you'll recognize social influences that may be operating. Reflection may not show you much about the social influences on your own thinking or behavior. But if you can see what social influences might be doing to *others*, it's a safe bet you might be susceptible as well.

Realize that situational factors usually influence your behavior and that of others more than they seem to, whereas dispositional factors are usually less influential than they seem. Don't assume that a given person's behavior in one or two situations is necessarily predictive of future behavior. And don't assume that the person has a trait or belief or preference that has produced the behavior.

Realize that other people think their behavior is more responsive to situational factors than you're inclined to think—and they're more likely to be right than you are. They almost certainly know their current situation—and their relevant personal history—better than you do.

Recognize that people can change. Since the time of the ancient Greeks, Westerners have believed that the world is largely static and that objects, including people, behave as they do because of their unalterable dispositions. East Asians have always thought that change is the only constant. Change the environment and you change the person. Later chapters argue that a belief in mutability is generally both more correct and more useful than a belief in stasis.

These injunctions can become part of the mental equipment you use to understand the world. Each application of the principles makes further applications more likely because you'll be able to see their utility and because the range of situations in which they can be applied will consequently increase.

3. The Rational Unconscious

We generally feel that we're fairly knowledgeable about what's going on in our heads—what it is we're thinking about and what thinking processes are going on. But an absolute gulf separates this belief from reality.

As should be clear from the two chapters you've just read, a huge amount of what influences our judgments and our behavior operates under cover of darkness. Stimuli we hardly take conscious note of—if we pay attention to them at all—can have marked effects on our behavior. Many of the stimuli that we do notice have consequences far beyond what seems plausible.

We don't know that we walk more slowly when we're thinking about elderly people. We don't know that we rated Jennifer's performance more highly than Jasmine's in part because we know that Jennifer has higher social class origins than Jasmine. We don't realize that, contrary to our usual voting behavior, we endorsed higher tax rates for education in our community in part because the vote took place in a school in this election. We don't realize that we signed Bob's petition but not Bill's in part because of the clearer font in Bob's petition. We don't realize that we found Marian to be a warmer person than Martha in part because we shared coffee with Marian and iced tea with Martha. Although it feels as if we have access to the workings of our minds, for the most part we don't. But we're quite agile in coming up with explanations for our judgments and behavior that bear little or no resemblance to the correct explanations. These facts about awareness and consciousness are laden with important implications for how we conduct our daily lives.

Consciousness and Confabulation

Many years ago, Timothy Wilson and I began a program to find out how people explain to themselves cognitive processes that influence their judgments in ordinary everyday situations.[1] We expected to find that when people lack a theory about what's going on in their heads, or hold to a wrong theory, they can misidentify what really was happening. And they do this because they have no window on cognitive processes—just theories about what those processes might be.

In one simple study we had people memorize word pairs. Then we asked them to participate in a word association study. For example, one of the word pairs in the first study was "ocean-moon." In the word association task in the "second study" we asked them to name a detergent. You probably won't be surprised to know that having memorized that particular word pair made it more likely that the detergent named would be "Tide." (Some participants, of course, were not exposed to the "ocean-moon" pair so that we would have a base for comparison.) After the word association task was over, we asked participants why they came up with the word that they did. They almost never mentioned the word pair they had learned. Instead, participants focused on some distinctive feature of the target ("Tide is the best-known detergent"), some personal meaning of it ("My mother uses Tide"), or an emotional reaction to it ("I like the Tide box").

When specifically asked about any possible effect of the word cues, approximately a third of the subjects did say that some of the words had probably had an effect. But there is no reason to assume that those participants were actually aware of the link. For some of the influential word pairs, not a single participant thought they had had an effect on their associations. For other pairs, many participants claimed there had been an influence of the word pairs, whereas in fact only very few had been influenced. (We know this because we know the extent to which learning the word pairs actually affected the probability of coming up with the target word.) This study establishes that not only can people fail to be aware of a process that went on in their heads, they can fail to retrieve that process when asked directly about it.

People can fail not merely to identify that some factor A influenced some outcome B, they may actually believe that it was outcome B that influenced factor A.

In some of our studies, participants' reports about the reasons for their judgments actually reversed the real causal direction. For example, we showed students an interview with a college teacher who spoke with a European accent. For half the participants the teacher presented himself as a warm, agreeable, and enthusiastic person. The other half of the participants saw the teacher present himself as a cold, autocratic martinet who was distrustful of his students. Participants then rated the teacher's likability and also three attributes that were by their nature essentially invariant across the two experimental conditions: his physical appearance, his mannerisms, and his accent.

Students who saw the warm teacher of course liked him much better than participants who saw the cold version of the teacher, and the students' ratings of his attributes showed they were subject to a very marked *halo effect*. A halo effect occurs when knowing something very good about a person (or very bad) colors all kinds of judgments about the person. The great majority of the participants who saw the warm version rated the teacher's appearance and mannerisms as attractive, and most were neutral about his accent. The great majority of the participants who saw the cold version rated all these qualities as unpleasant and irritating.

Were the participants who saw the friendly version of the teacher aware that their positive feelings for him had influenced their ratings of his attributes? And were those who saw the cold version aware that their negative feelings had influenced their ratings of his attributes? We asked this question of some of the participants. They strongly denied any effect of their positive or negative feelings for the teacher on their ratings of his attributes. (In effect, "Give me a break, of course I can make a judgment about someone's accent without being influenced by how much I like him.") We asked other participants the reverse question—how much did their feelings about the teacher's attributes influence their overall liking for him? Participants who saw the warm version denied that their feelings about the teacher's attributes influenced their overall evaluation of him. But participants who saw the cold version felt that their dislike of each of the three attributes had probably contributed to their like of him. So those participants got things exactly backward. Their dislike of the teacher had lowered their evaluation of his appearance, his manner-isms, and his accent, but they denied such an influence and maintained instead that their dislike of these attributes had decreased their overall liking of him!

So we can be confident that we have not been influenced by some-thing that did in fact influence us, and we can be equally confident that something that did not influence us did have an effect. This degree of confusion can wreak havoc on our decisions about people. We don't always know why we like them or dislike them, and therefore can make serious mistakes in dealing with them, trying, for example, to get them to change attributes and behavior that we think are making us dislike them but that are in fact neutral and have nothing to do with our overall feelings about them.

Subliminal Perception and Subliminal Persuasion

People needn't be aware of a stimulus at all in order for it to affect them. The term "subliminal" is used to refer to a stimulus that a person is not consciously aware of. (A limen is the point at which a stimulus such as a light, noise, or occurrence of any kind becomes detectable.)

A famous finding in psychology is the discovery that the more times people are exposed to a stimulus of a given type—ditties, Chinese char-acters, Turkish words, people's faces—the more people like the stimulus (so long as they don't dislike the stimulus to begin with).[2] This so-called *mere familiarity effect* is shown by a study in which people listen to a communication played for one ear while having various tone sequences piped into the other ear. It turns out that the more frequently people hear a given tone sequence, the more they like it. And this is true even when people have no awareness that the tones were played for them and no ability after the experiment is over even to distinguish tone sequences that had been played for them many times from those they had never heard.

The psychologists John Bargh and Paula Pietromonaco presented words on a computer screen for one-tenth of a second, and then to make sure the participants were unaware of what they had seen, they presented a "masking stimulus" consisting of a line of Xs where the word had been.[3] Some participants were exposed to words with a hostile meaning and some to neutral words. The participants then read about "Donald," whose behavior could be construed either as hostile or as merely neutral. ("A salesman knocked at the door, but Donald refused to let him enter.") Participants exposed to the hostility-related words rated Donald as being more hostile than did participants exposed to the neutral words. Immediately after reading the paragraph, participants couldn't distinguish words they had seen from those they hadn't, and didn't even know that words had been flashed at all.

Findings such as these raise a question as to whether there is such a thing as subliminal persuasion—being influenced to believe something or do something in response to a stimulus presented at such a low intensity that people can't report whether they have seen anything. There's been quite a bit of research on this topic over the years, but very little of it was well enough conducted to be convincing one way or the other.

Some recent marketing research indicates that subliminal stimuli can in fact influence product choice. For example, if you make people thirsty and then expose them to a particular brand name presented so briefly that they're unaware of it, they're more likely to choose that brand when given a choice between it and a brand that wasn't presented.[4]

But there's no question that stimuli that are *supraliminal* (above the level of awareness)—but seemingly incidental and little noticed—can have an effect on consumer choices.[5] Even so trivial a stimulus as the color of the pen someone uses to indicate product choice can be influential.[6] People writing with an orange pen choose more orange products in a consumer survey than people writing with a green pen. Contextual cues matter for consumer choice as for everything else.

How to Perceive Before Perceiving

In the popular mind, the unconscious is primarily the repository of repressed thoughts about violence and sex and other things best unmentioned. In fact, however, the conscious pot has no right to call the unconscious kettle black. There's plenty of sex and violence roaming around in the conscious mind. If you give a buzzer to college students and have them write down what they were thinking about each time the buzzer goes off, much of the time it's a sexual thought. And the great majority of college students report that they've entertained thoughts of killing someone.[7]

Rather than just mucking about with unacceptable thoughts, the unconscious mind is constantly doing things that are useful—even indispensable.

The unconscious mind "preperceives" for us. Think of our perceptual systems as monitoring unconsciously a vast array of stimuli. The conscious mind is aware of only a small fraction of what's in that array. The unconscious mind forwards to the conscious mind those stimuli that will interest you or that you need to deal with.

If you doubt this claim, think of the situation of being in a room with a grandfather clock. You've been listening to its ticking, whether you know it or not. How can we be sure of that? Because if the clock stops ticking, you instantly notice that. Or consider the "cocktail party phenomenon." You're standing in a room with thirty other people straining to hear the person you're talking to over the hubbub. You're hearing nothing but what she is saying. But no, actually, you've been hearing a lot else besides. If someone five feet away from you mentions your name, you instantly pick that up and orient toward the speaker.

Just as the unconscious mind has a much larger perceptual capacity than the conscious mind, it has a far greater ability to hold multiple elements in thought and a far greater range of *kinds* of elements that can be held in thought. A consequence of this is that the conscious mind can mess up your evaluation of things if you let it get into the act. If you're encouraged to verbally express your reactions to objects such as art posters or jams and tell what you like and dislike about each one, your choices are likely to be worse than if you simply think about the objects for a while and then make a choice.[8] We know the judgments are worse because people asked to verbalize their thought processes report being less satisfied with the object they chose when they're asked to rate it at some later point.

Part of the reason conscious consideration of choices can lead us astray is that it tends to focus exclusively on features that can be verbally described. And typically those are only some of the most important features of objects. The unconscious considers what can't be verbalized as well as what can, and as a result makes better choices.

If you cut the conscious mind out of the process of choosing, you can sometimes get better results. In a study supporting this conclusion, Dutch investigators asked students to pick the best of four apartments. Each apartment had some attractive features ("very nice area of town") and some unattractive features ("unfriendly landlord").[9] One apartment was objectively superior to the others because it had eight positive, four negative, and three neutral features—a better mix than the others. Some participants had to make their choice immediately, with little time to think about the choice either consciously or nonconsciously. Other participants were asked to think carefully about their choice for three minutes and review all the information as best they could. These participants had plenty of time for conscious consideration of the choice. A third group saw the same information as the others, but participants weren't able to process it consciously because they had to work on a very difficult task for the three-minute period. If they were processing the information about the apartments, they were doing so without awareness.

Remarkably, participants in this last, distracted group working on the difficult task were almost a third more likely to pick the right apartment than the group allowed plenty of time for conscious thought. Moreover, the latter group failed to make better choices than the group given scarcely any time to think.[10] These findings obviously have profound relevance for how we should make choices and decisions in life. We'll have occasion to be reminded of this in the next part of the book, where we discuss theories of how people make choices and how they can maximize the likelihood that those choices will be the best possible.

Learning

The unconscious mind can actually be superior to the conscious mind in learning highly complex patterns. More than that, in fact: it can learn things that the conscious mind can't. Pawel Lewicki and his coworkers asked people to pay attention to a computer screen divided into four quadrants.[11] An X would appear in one of the quadrants. The participant's task was to press a button predicting which quadrant the X was going to appear in. Though participants didn't know it, whether an X appeared in a given quadrant was dictated by a very complicated set of rules. For example, an X never appeared twice in a row in the same quadrant, an

X never returned to its original location until it had appeared in at least two of the other quadrants, an X in the second location determined the location of the third, and the fourth location was determined by the location on the previous two trials. Could people learn such a complicated rule system?

Yes. We know people can learn them because (1) participants became faster over time at pressing the correct button and (2) when the rules suddenly changed, their performance deteriorated badly. But the conscious mind was not let in on what was happening. Participants didn't even consciously recognize that there *was* a pattern, let alone know exactly what it was.

Participants were adept, however, at accounting for suddenly worsened performance. That may have been especially true because the participants were psychology professors (who incidentally knew they were in a study on nonconscious learning). Three of the professors said they had just "lost the rhythm." Two accused the experimenter of putting distracting subliminal messages on the screen.

Why don't we recognize consciously just what pattern it is that we've learned? I'll ask the curt question, "Why should we?" For most purposes, what's crucial is that we learn a pattern, not that we be able to articulate exactly what the rules behind the pattern are.

The unconscious mind is very good at detecting all kinds of patterns. Imagine a computer grid with one thousand pixels that can be either black or white. Take half of that grid and randomly make some proportion of the pixels black and some white. Then flip the half-grid over and create the mirror image of the original. Place the two images side by side. You will instantly see the symmetry between the two halves. How is it that you see there is perfect symmetry? It's certainly not by conscious calculation, determining whether each pixel in the mirror-image location is the same or not. The number of calculations necessary to determine whether there is perfect symmetry is five hundred thousand. That computational trick couldn't be performed quickly even by computers until relatively recently.

Laborious calculation is clearly not involved in complex pattern detection. Seeing a mirror image is instantaneous and automatic. If it's there, you can't *not* see it. And if someone were to ask you what the pattern of pixels was exactly, you would be utterly stumped (unless by some miracle the pixels formed themselves into a few clear and readily describable shapes). Your nervous system is an exquisitely designed pattern detector. But the process by which it sees patterns is completely opaque to us.

Unfortunately, we're a little too good at detecting patterns. We see them even when they aren't there. As we'll see in Part III, we're often confident that a collection of events that are utterly random have been caused by some agent such as another person.

Problem Solving

Prime numbers are those that are divisible only by 1 and themselves. Euclid proved more than two thousand years ago that there are an infinite number of prime numbers. An interesting fact about prime numbers is that they often appear as "twins" differing only by 2—such as 3 and 5, 17 and 19. Are there an infinite number of twin primes? This problem has fascinated eminent mathematicians and amateurs alike but no solution appeared for the past two millennia and more. Computers have discovered twin pairs as large as $3,756,801,695,685 \times 2^{666,689} -1$. But brute computing power could never establish the truth of the conjecture, and a solution to the twin prime problem has long been a mathematical holy grail.

On April 17, 2012, the *Annals of Mathematics* received a paper from an obscure mathematician at the University of New Hampshire that claimed a giant leap toward verifying the twin primes conjecture.[12] The author was fiftysomething Yitang Zhang, who had spent many years adrift in jobs such as accountant and even Subway employee before he finally got a job at UNH.

Mathematics journals are constantly fielding grandiose claims from obscure mathematicians, but the editors at the *Annals* found Zhang's arguments plausible on the surface and promptly sent the paper out for review. Three weeks after receipt of the paper by the *Annals*—warp speed by academic standards—all the referees pronounced the claims valid.

What Zhang proved was that there are infinitely many pairs of prime numbers that differ by 70 million or less. No matter how far you go into the region of spectacularly large prime numbers, and no matter how infrequent they become, you will keep finding prime pairs that differ by less than 70 million.

Number theorists pronounced the result "astounding." At the invitation of Harvard University, Zhang gave a lecture on his work to a huge crowd of Cambridge academics. His talk impressed his hearers as much as the paper had wowed reviewers.

Zhang had worked on the twin prime conjecture for three years, making no progress whatsoever. Then the solution suddenly came to him, not while he was toiling away on the problem in his office, but while he was sitting in a friend's backyard in Colorado while he waited to leave for a concert. "I immediately knew that it would work," he said.

Now that the unconscious had done its part, the hard conscious work began. It took Zhang several months to work out all the details of the solution.

Zhang's experience is quite typical of creative problem solving at the very highest level. There's a striking uniformity in the way creative people—artists, writers, mathematicians, and scientists—speak about how they created their products. The American poet Brewster Ghiselin collected into one volume a number of essays on the creative process by a variety of highly inventive people from Poincaré to Picasso.[13]

"Production by a process of purely conscious calculation seems never to occur," Ghiselin says. Instead, his essayists describe themselves almost as bystanders, differing from observers only in that they are the first to witness the fruits of a problem-solving process that's hidden from conscious view.

Ghiselin's essayists insist that (a) they had little or no idea what factors prompted the solution, and (b) even the fact that thought of any kind about the problem was taking place is sometimes unknown.

The mathematician Jacques Hadamard reports that "on being very abruptly awakened by an external noise, a solution long searched for appeared to me at once without the slightest instant of reflection on my part . . . and in a quite different direction from any of those which I previously tried to follow." The mathematician Henri Poincaré records that "the changes of travel made me forget my mathematical work . . . At the moment when I put my foot on the step [of the omnibus] the idea came to me, without anything in my former thoughts seeming to have paved the way for it, that the transformations I had used to define the Fuchsian functions were identical with those of non-Euclidean geometry." The philosopher and mathematician Alfred North Whitehead wrote of "the state of imaginative muddled suspense which precedes successful inductive generalization."

The poet Stephen Spender describes "a dim cloud of an idea which I feel must be condensed into a shower of words." The poet Amy Lowell wrote, "An idea will come into my head for no apparent reason; 'The Bronze Horses,' for instance. I registered the horses as a good subject for a poem; and, having so registered them, I consciously thought no more about the matter. But what I had really done was to drop my subject into the subconscious, much as one drops a letter into the mail-box. Six months later, the words of the poem began to come into my head, the poem—to use my private vocabulary—was 'there.'"

What's true for the most creative people in history working on the most interesting ideas is true for you and me working on much more mundane problems.

The better part of a century ago, the psychologist N.R.F. Maier showed people two cords hanging from the ceiling of a laboratory strewn with many objects such as clamps, pliers, and extension cords.[14] Maier told the participants that their task was to tie the two ends of the cords together. The difficulty was that the cords were placed far enough apart that the participants couldn't reach one while holding on to the other. Maier's participants quickly came up with several of the solutions, for example, tying an extension cord to one of the ceiling cords. After each solution, Maier told the participants, "Now do it a different way."

One of the solutions was much more difficult than the others, and most participants couldn't discover it on their own. While the participant stood perplexed, Maier would be wandering around the room. After the participant had been stumped for several minutes, Maier would casually put one of the cords in motion. Then, typically within forty-five seconds of this clue, the subject picked up a weight, tied it to the end of one of the cords, set it to swinging like a pendulum, ran to the other cord, grabbed it, and waited for the first cord to swing close enough that it could be seized. Maier immediately asked the participants to tell how they thought of the idea of a pendulum. This question elicited such answers as, "It just dawned on me," "It was the only thing left," "I just realized the cord would swing if I fastened a weight to it."

A psychology professor participant gave a particularly rich account:

"Having exhausted everything else, the next thing was to swing it. I thought of the situation of swinging across a river. I had imagery of monkeys swinging from trees. This imagery appeared simultaneously with the solution. The idea appeared complete."

After hearing their explanations, Maier grilled his participants about whether the swinging cord had had any effect on them. Almost a third allowed that it had. But there's no reason to believe that those participants were actually aware of the role of the cord. Rather, they might simply have found the theory plausible and endorsed it. To make sure that participants had no genuine introspective awareness of their thinking, Maier conducted a new study in which he twirled a weight on a cord. This hint was useless; no one solved the problem after this cue was provided. For other participants, Maier twirled the weight and then shortly afterward swung the cord. Most participants then promptly applied the pendulum solution. Upon being questioned, however, all of these participants insisted that the twirling weight had helped them to solve the problem and denied that swinging the cord had had any effect!

The lesson of Maier's experiment is profound. Problem-solving processes can be as inaccessible to consciousness as any other kind of cognitive process.

Why Do We Have Conscious Minds Anyway?

The most important thing to know about the unconscious is that it's terrific at solving certain kinds of problems that the conscious mind handles poorly if at all. But although the unconscious mind can compose a symphony and solve a mathematical problem that's been around for centuries, it can't multiply 173 by 19. Ask yourself to figure that out as you drift off to sleep and see if the product pops into your mind while you're brushing your teeth the next morning. It won't.

So there's a class of rules—probably a very large class of even simple rules like those of multiplication—that the unconscious mind can't operate with. (Yours and mine, that is. Savants can do it somehow.) It seems paradoxical in the extreme that an operation any fourth grader can carry out consciously couldn't be handled by a von Neumann unconsciously. The unconscious mind operates according to rules, for sure. But we don't really have any good way as yet to characterize just which rule systems require consciousness and which can operate unconsciously—or whether there are any that can operate both ways.

We do know that a given task can be carried out using either conscious rules or unconscious ones. But the solutions yielded by the one can be, and perhaps usually are, utterly different. Herbert Simon, the Nobel Prize–winning economist–computer scientist–psychologist–political scientist, attacked the contentions by Tim Wilson and me that there was no such thing as conscious observation of mental processes. He had found that people who were solving problems while thinking aloud could describe accurately the processes by which they solved them. But his examples only showed that people were capable of generating theories about what rules they were using to solve the problems and that these theories were sometimes accurate—not at all the same thing as observing the processes.

In conscious problem solving we're aware of: (1) certain thoughts and perceptions that are in our heads, (2) particular rules that we believe govern (or should govern) how we deal with those thoughts and perceptions, and (3) many of the cognitive and behavioral outputs of whatever mental processes are going on. I know the rules of multiplication, I know the numbers 173 and 19 are in my head, I know I must multiply 3 by 9, save the 7 and carry the 2, and so forth. I can check that what's available to my consciousness is consistent with the rules that I know to be appropriate. *But none of this can be taken to mean that I am aware of the process by which multiplication is carried out.*

In conversation, Simon actually gave me the perfect example of how a given task can be carried out operating either by unconscious rules or by rules that are represented consciously.

When people first play chess, they move the pieces around without being able to tell you what rules, if any, they're following. But they are indeed following rules. Their technique is called "duffer strategy"—whose rules are well known to experts.

Later, if people stick with chess for a while—read books on it and talk to highly competent players—they play according to rules that are quite conscious and that they can describe accurately. But I would insist they can't *see* what's going on; they can simply check that their behavior is consonant with the consciously represented rules and with the thoughts they have while using those rules.

It's unfortunate that we can't monitor the processes that underlie the solutions to complex problems. But it's even more unfortunate that we often are convinced that we can. It can be hard to change someone's mind about the wisdom of some strategy or tactic when the person is dead certain that he knows just what's going on and he's not making the mistake you're trying to point out to him.

When players become genuinely expert, they once again can no longer accurately describe the rules they're using. This is partly because they no longer have conscious representation of many of the rules they learned as an intermediate player and partly because they have induced unconsciously the strategies that made them masters or grandmasters.

The assertion that we have no access to the processes that underlie our judgments may not seem so radical in light of two considerations.

1. We claim to know the processes that underlie judgment and behavior, but we make no claim that we have awareness of the processes underlying perception or retrieval of information from memory. We know the latter processes are completely beyond our ken. Perfectly adequate processes producing perception and memory take place without our awareness. Why should cognitive processes be any different?

2. From an evolutionary standpoint, why would it be important to have access to the mental processes that are doing the work for us? The conscious mind has enough to do without also having to be aware of mental processes that are producing the needed inferences and behavior.

To say that there's no direct awareness of mental processes is not to say that we're usually wrong about what goes on behind the scenes. Often, maybe usually, I can say with justified confidence what were the most important stimuli I was attending to, and why I behaved as I did. I know that I swerved the car to avoid hitting the squirrel. I know that the main reason I gave at the office was because everyone else was making a donation. I know that I was anxious about the exam because I hadn't studied very much.

But in order to be right about what drives my judgments and behavior, I have to have a correct theory. I have no theory that says I'm less likely to cheat if there's a picture of a coconut hanging over the honest box, or that voting in a church has made me more likely to vote against abortion. Or that hunger is making me unsympathetic to a job applicant. Or that fishy smells make me doubt what I'm reading. Or that holding a hot coffee cup makes me think you're a warm person. Indeed, what would a theory of such things look like? Anything less general and less useless than "Who knows what all is going on to affect my behavior?"

If we had theories about the processes underlying those behaviors, we would draw on them as reasons for behaving as we do. In many cases, in fact, we would resist those processes and often produce a better outcome. But, lacking the correct theories about these kinds of processes, we can't have correct explanations for why we behave as we do.

Summing Up

This chapter has many implications for how we should function in daily life. Here are a few of the most important.

Don't assume that you know why you think what you think or do what you do. We don't know what may have been the role played by little-noticed and promptly forgotten incidental factors. Moreover, we often can't even be sure of the role played by factors that are highly salient. Why should you give up belief in self-knowledge, and do so at the cost of self-confidence? Because you're less likely to do something that's not in your best interest if you have a healthy skepticism about whether you know what you really think or why you really do the things you do.

Don't assume that other people's accounts of their reasons or motives are any more likely to be right than are your accounts of your own reasons or motives. I frequently find myself telling other people why I did something. When I do that I'm often acutely aware that I'm making this up as I go along and that anything I say should be taken with more than a grain of salt. But my hearers usually nod and seem to believe everything I say. (With psychologists I usually have the grace to remind them there is no particular reason to believe me. Don't try that with nonpsychologists.)

But despite my recognition that my explanations are somewhere between "probably true" and "God only knows," I tend to swallow other people's explanations hook, line, and sinker. Sometimes I do realize that the person is fabricating plausible explanations rather than reporting accurately, but more typically I'm as much taken in as other people are taken in by my explanations. I really can't tell you why I remain so gullible, but that doesn't prevent me from telling *you* to carry a saltshaker around with you.

The injunction to doubt what people say about the causes of their judgments and behavior, incidentally, is spreading to the field of law. Increasingly it's recognized that what witnesses, defendants, and jurors say about why they did what they did or reached the conclusions that they came to are not to be trusted—even when they are doing their level best to be perfectly honest.[15]

You have to help the unconscious help you. Mozart seems to have secreted music unbidden. (And if you saw the movie *Amadeus*, you know that he frequently wrote down the output without ever blotting a note.) But for ordinary mortals, creative problem solving seems to require consciousness at two junctures.

1. Consciousness seems to be essential for identifying the elements of a problem, and for producing rough sketches of what a solution would look like. The *New Yorker* writer John McPhee has said that he has to begin a draft, no matter how crummy it is, before the real work on the paper can begin. "Without the drafted version—if it did not exist—you obviously would not be thinking of things that would improve it. In short, you may be actually writing only two or three hours a day, but your mind, in one way or another, is working on it twenty-four hours a day—yes, while you sleep—but only if some sort of draft or earlier version already exists. Until it exists, writing has not really begun" (McPhee, 2013). Another good way to kick the process off, McPhee says, is to write a letter to your mother telling her what you're going to write about.

2. Consciousness is necessary for checking and elaborating on conclusions reached by the unconscious mind. The same mathematicians who say that a given solution hit them out of the blue will tell you that making sure the solution was correct took hundreds of hours of conscious work.

The most important thing I have to tell you—in this whole book—is that you should never fail to take advantage of the free labor of the unconscious mind.

I teach seminars by posting a list of thought questions to serve as the basis for discussion for the next class. If I wait until the last minute to come up with those questions, it's going to take me a long time and the questions won't be very good. It's tremendously helpful for me to sit down two or three days before my deadline—just for a few minutes—and think about what the important questions might be. When I later start to work on the questions in earnest, I typically feel as if I'm taking the questions by dictation rather than creating them. If you're a student, a question for you: When is the right time to begin working on a term paper due the last day of class? Answer: The first day of class.

If you're not making progress on a problem, drop it and turn to something else. Hand the problem off to the unconscious to take a shot at it. When I used to do calculus homework, there would always come a point when I hit a problem that I absolutely could make no progress on. I would stew over the problem for a long time, then move on in a demoralized state to the next problem, which was typically harder than the previous one. There would follow more agonized conscious thought until I shut the book in despair. Contrast this with how a friend tells me that he used to deal with the situation of being stumped on a calculus problem. He would simply go to bed and return to the problem the next morning. As often as not the right direction to go popped into his head. If only I had known this person when I was in college.

I hope that having a clearer understanding of how your mind works will make it easier to understand how useful the concepts in this book can be. The fact that it may seem to you that it's unlikely that a given concept would be helpful doesn't mean you wouldn't use it—and use it properly—if you knew it. And the more you use a given concept, the less aware of using it you will become.

PART II

THE FORMERLY DISMAL SCIENCE

When you think of economists, the picture that likely comes to mind is a professor or government employee or corporation executive working out equations describing gross domestic product for various countries, predicting what the market will be the next year for coal, or advising the Federal Reserve about how to set rates for overnight loans. Work on such a large scale is called *macroeconomics*. Economists who do that kind of work are not getting as much respect lately as in times past. We have it on the authority of the Nobelist Paul Krugman that no economist predicted the Great Recession of 2008. (Except for one who had successfully predicted nine of the previous five recessions!) Indeed, some critics claim that erroneous mathematical models by economists in investment banks and the ratings companies contributed to the circumstances that made the recession possible.

A pair of economists won the Nobel Prize in 2013 for showing that the stock and bond markets are utterly accurate and rational. Stocks and bonds are always worth exactly what they are selling for at any given moment; consequently it's impossible to beat the market by trying to time it. Another economist won the Nobel the same year for showing that markets are less than completely rational and driven in part by emotional overreactions; consequently it's quite possible to make money by timing the market successfully! (I have economist friends who tell me these positions aren't really contradictory. I just pass that along.)

Regardless of which economists are right about the big-picture issues, it's not likely that you need to know much about macroeconomics in order to live your life in the most effective way. There is another branch of economics, however, that *is* concerned with how to live your life. *Microeconomics* is the study of the way individuals, corporations, and entire societies make choices. Microeconomists are also in the habit of telling us how we *should* make decisions. Both *descriptive* micro-economics and *prescriptive* microeconomics are embroiled in contro-versy. Over the last hundred years or so, many different descriptive theories of choice and many prescriptive theories of choice have been proposed. The field has been near agreement on these matters from time to time, but then someone comes along with a new paradigm and battles begin anew.

The most recent microeconomic warfare has resulted from cogni-tive psychologists and social psychologists entering the fray. The field of *behavioral economics* is an amalgam of psychological theories and re-search and novel economic perspectives. This hybrid seeks to overturn traditional descriptive and prescriptive theories of choice. And behavioral economists are beginning to move into the business of helping people to make choices. They're not only telling you how to make choices, they're engineering the world so that you make choices they believe to be optimal. If this sounds Orwellian, it really isn't. The tongue-in-cheek name that some behavioral economists use to describe their enterprise is "libertarian paternalism." These economists will tell you how to make choices and arrange the world so that you'll be likely to make good ones. But they're not forcing you. You can always choose to ignore the choices they steer you toward making.

As you might expect, the entry of psychologists onto the economic scene has brought along some of the basic assumptions discussed in the previous chapters. These include the contention that we don't always know why we make the choices that we make, and that our choices, like other behaviors, are not always fully rational. That's why you need some help, according to behavioral economists.

Chapter 4 presents some fairly traditional economic theory about how people make choices and how they should make them. Most of the material there is accepted by most economists, including those of the maverick behavioral sort. Chapter 5 shows the kinds of errors that people can make across the entire spectrum of daily choices. Knowing about those errors will improve how you approach the innumerable choices you face every day. Chapter 6 presents the behavioral economic view of how we make choices, how we should make choices, and why it's a good idea for experts to nudge you in the direction of superior choices.

4. Should You Think Like an Economist?

When difficult cases [decisions] occur, they are difficult chiefly because while we have them under consideration, all the reasons pro and con are not present to the mind at the same time . . . To get over this, my way is to divide half a sheet of paper by a line into two columns; writing over the one "Pro," and the other "Con." Then . . . I put down under the different heads short hints of the different motives . . . for and against the measure . . . I endeavor to estimate their respective weights; where I find one on each side that seem equal, I strike them both out. If I find a reason pro equal to two reasons con, I strike out three . . . and thus proceeding I find at length where the balance lies . . . And, though the weight of reasons cannot be taken with the precision of algebraic quantities, yet when each is thus considered, separately and comparatively, and the whole lies before me, I think I can judge better, and am less liable to take a rash step.

—Benjamin Franklin

Benjamin Franklin's suggestions about how to deal with choice are what we now call *decision analysis*. His procedure is a more detailed account of a method for decision making initially proposed in the middle of the seventeenth century by the mathematician, physicist, inventor, and Christian philosopher Blaise Pascal. In carrying out what is called *expected value analysis*, you list the possible outcomes of each of a set of choices, determine their value (positive or negative), and calculate the probability of each outcome. You then multiply value by probability. The product gives you the expected value of each course of action. You then pick the action with the highest expected value.

Pascal described his decision theory in the context of considering his famous wager: Everyone has to decide whether to believe in God or not. At the heart of his analysis was what we would call today a *payoff matrix*:

TABLE 1. PAYOFF MATRIX FOR PASCAL'S WAGER

	God Exists	God Does Not Exist
Belief in God	+∞ (infinite gain)	−1 (finite loss)
Disbelief in God	−∞ (infinite loss)	+1 (finite gain)

If God exists and we believe in him, the reward is eternal life. If he exists and we do not believe in him, the consequence is eternal damnation. If God does not exist and we believe in him, there is a loss that is not too substantial—mostly forgoing guilty pleasures and avoiding selfish behavior that harms others. If God does not exist and we disbelieve, there is a relatively minor gain—indulging those guilty pleasures and behaving selfishly. (I note parenthetically that many psychologists today would say that Pascal may have gotten the finite gains and losses reversed. It actually is better for your well-being to give money than to receive it,[1] and kind consideration of others makes one happier.[2] But this doesn't affect the logic of Pascal's payoff matrix.)

Pity the poor atheist if Pascal got the payoffs right in the event that God exists. Only a fool would fail to believe. But unfortunately you can't just grunt and produce belief.

Pascal had a solution to this problem, though. And in solving the problem he invented a new psychological theory—what we would now call cognitive dissonance theory. If our beliefs are incongruent with our behavior, something has to change: either our beliefs or our behavior. We don't have direct control over our beliefs but we do have control over our behavior. And because dissonance is a noxious state, our beliefs move into line with our behavior.

Pascal's prescription for atheists is to proceed "by doing everything as if they believed, by taking holy water, by having Masses said, etc. . . . This will make you believe . . . What have you to lose?"

Social psychologists would say that Pascal got it just right. Change people's behavior and their hearts and minds will follow. And his decision theory is basically the one at the core of all subsequent normative decision theories.

Cost-Benefit Analysis

An economist would maintain that, for decisions of any consequence at all, you have to conduct a *cost-benefit analysis*, which is a way of calculating expected value. The formal definition of cost-benefit analysis is that the action that has the greatest net benefit—benefit minus cost—should be chosen from the set of possible actions. More specifically, one should do the following.

1. List alternative actions.
2. Identify affected parties.
3. Identify costs and benefits for each party.
4. Pick your form of measurement (which will usually be money).
5. Predict the outcome for each cost and benefit over the relevant time period.
6. Weight these outcome predictions by their probability.
7. Discount the outcome predictions by a decreasing amount over time (a new house is worth less to you twenty years from now than it is now because you have less time left in your life to enjoy it). The result of the discounting is the "net present value."
8. Perform a sensitivity analysis, meaning one adjusts the outcome of cost-benefit analysis due to, for example, possible mistakes in estimating the costs and benefits or errors in estimating probabilities.

Needless to say, all this sounds daunting, and it actually leaves out or simplifies some steps.

In practice, a cost-benefit analysis can be considerably less complicated than what's implied by the above list. An appliance company might need to decide whether to put out either one or two colors of its new juicer; an auto company might need to decide between two versions of an auto model. The costs and benefits are easy to identify (though estimating probabilities for them can be very difficult), money is the obvious measure, the discount rate is the same for both options, and the sensitivity analysis is relatively easy to perform.

Decisions by individuals can be similarly uncomplicated. Let's consider a real one confronted by a couple who are friends of mine. Their old refrigerator is on its last legs. Choice A is to buy an ordinary refrigerator like most people have, costing in the range of $1,500 to $3,000, depending on quality and features such as ice maker and water cooler. Such refrigerators have some unattractive features, a repair record that's not great, and a relatively short life expectancy—perhaps ten to fifteen years. Choice B is to buy a qualitatively different kind of refrigerator that is extremely well built and has many attractive features. It functions beautifully, its repair record is excellent, and it can be expected to last in the range of twenty to thirty years. But it costs several times as much as an ordinary refrigerator.

Calculating expected value is not terribly difficult in such a case. Benefits and costs are pretty clear, and it's not all that difficult to assign probabilities to them. Though the choice might have been difficult for them, my friends can feel comfortable about their decision because they had considered everything they ought to have, and they had assigned reasonable values for costs and benefits and for the probabilities of those costs and benefits.

But consider a somewhat more difficult choice involving assessment of multiple costs and benefits. You're considering buying either a Honda or a Toyota. You do not—or should not—buy a Honda whose assets, taken as a whole, are of value X rather than a Toyota whose somewhat different assets, taken as a whole, are also of value X—if the Honda is more expensive.

Well, of course. But the devil is in the details.

Problem number one is how to limit the *choice space*—the options you're going to actually consider. Who said you should be choosing between a Honda and a Toyota? How about a Mazda? And why stick with Japanese cars? Volkswagens are nice and so are Fords.

Problem number two is when to stop searching for information. Did you really look at every aspect of Hondas and Toyotas? Do you know the expected gasoline consumption per year? The relative trade-in values of the two cars? The capacity of the trunk? The *optimizing* choice— making the best decision possible—is not a realistic goal for many real-world decisions. If we really tried to optimize choice, we would be in the position of the philosopher's donkey starving between two bales of hay. ("This one looks a little fresher. Looks like there's more hay in that one. This one is a little closer.")

Enter the economist–political scientist–psychologist–computer scientist–management theorist introduced in the previous chapter, namely Herbert Simon. He attempted to solve these two problems with cost-benefit theory. It's often not rational, he says, to try to optimize choice. It's the thing to do for a high-speed computer with infinite information, but not for us mortals. Instead, our decision making is characterized by *bounded rationality*. We don't seek to optimize our decisions; rather we *satisfice* (the word is a compound of "satisfy" and "suffice"). We should spend time and energy on a decision in proportion to its importance. This amendment to standard microeconomic theory is surely correct as far as it goes, and Simon won the Nobel Prize in economics for this principle. People who spend ten minutes deciding between chocolate and vanilla are in need of help. And, on the other hand, "Marry in haste, repent at leisure."

But there's a problem with the concept of satisficing. It's fine as a *normative prescription* (what you should do), but it's really not a very good description of the way people actually behave. They may spend more time shopping for a shirt than for a refrigerator and exert more energy pricing barbecue grills than shopping for a mortgage rate.

For a spectacular example of poor calibration of choice time in relation to choice importance, consider that the most important financial decision that most academics ever make takes them about two minutes. When they go to fill out employment papers, the office administrator asks how they want to allocate their retirement investments between stocks and bonds. The new employee typically asks, "What do most people do?" The reply comes back, "Most people do fifty-fifty." "Then that's what I'll do." Over the past seventy years or so, that decision would have resulted in the professor netting substantially less money at retirement than he would have with a decision for a 100 percent stock allocation. (But remember, I'm not a real financial analyst. And if you're going to follow my advice despite my lack of expertise, do remember that some analysts advise that a few years before retirement, you should take a considerable amount out of stocks and put it into bonds and cash so that you suffer less damage should the stock market be in a trough at retirement time.)

So what's a reasonable amount of time to spend on a decision to buy a car? Of course, what's reasonable differs from person to person. Rich people don't have to worry about which options they should choose. Just get 'em all! And if rich people have bad outcomes because they hadn't calculated probabilities correctly, they can just throw some money at the problem. But for most people a few hours or even days given over to research on automobiles seems sensible.

Now consider an extremely complicated and consequential choice. Here's a real one that was confronting a friend at the time of this writing.

My friend, who is a professor at a university in the Midwest, was recently made an offer by a university in the Southwest. The university wanted my friend to start a center for the study of a field of medicine that my friend had cofounded. No such center existed anywhere in the world, and medical students and postdoctoral fellows had no place to go to study the field. My friend is eager for there to be such a center and would very much like to put his stamp on it.

Here's a partial list of the costs and benefits he had to calculate.

1. Alternative actions were easy: go or stay.
2. Affected parties: my friend, his wife, their grown children, both of whom lived in the Midwest, potential undergraduate students, medical students and postdocs, the world's people at large—since there are considerable medical implications of any findings in my friend's field and it's possible that there would be more such findings if there were a center devoted to this field.
3. Identifying costs and benefits for my friend and his wife were a mixed bag. Some of the benefits were easy to identify: the excitement of starting a new center and advancing his field, escaping midwestern winters, a higher salary, a change of intellectual scenery. Assessing the probabilities of some of those things, not so easy. Some of the costs were equally clear: the hassles of moving, the burdens of administration, southwestern summers, leaving treasured friends and colleagues. But impact on the world? Very difficult to contemplate: no way to

know what the findings might be or even how much more likely they might be if my friend, rather than someone else, took the helm of the center. Benefits and costs to my friend's wife were fewer to calculate because her occupation as a novelist was portable and wouldn't change, but values and probabilities were difficult to estimate for her as well.

4. Measurement? Money works for salary. But how much is it worth to have a sunny January day with a high of sixty degrees versus a cloudy January day with a high of twenty? How much is the estimated excitement and pleasure of setting up a center offset by the aggravation of trying to recruit staff and administer the center? How about the benefits and costs (monetary and otherwise) for discoveries as yet unknown? Hopeless.

5. Discounting? Works well for salary, but difficult to impossible for most of the rest.

6. Perform a sensitivity analysis? What to say other than that the possible range of values for most of the benefits and costs is very large?

So why do the cost-benefit analysis at all, since there are so many imponderables?

Because, as Franklin said, your judgment will be better informed and you will be less likely to make a rash decision. But we shouldn't kid ourselves that the exercise is always going to come out with a number that will tell us what to do.

A friend of mine once carried out a cost-benefit analysis for an important move she was considering making. As she was nearing the end of the task, she thought to herself, "Damn, it's not working out right! I'll have to get some pluses on the other side." And there was her answer. As Pascal said: "The heart has its reasons of which reason knows nothing." And as Freud said, "When making a decision of minor importance, I have always found it advantageous to consider all the pros and cons. In vital matters, however . . . the decision should come from the unconscious, from somewhere within ourselves."

My friend's heart quite properly overruled her head, but it's important to be aware of the fact that the heart is also influenced by information. As I pointed out in the previous chapter, the unconscious needs all possible relevant information, and some of this information will be generated only by conscious processes. Consciously acquired information can then be added to unconscious information, and the unconscious will then calculate an answer that it delivers to the conscious mind. Do by all means perform your cost-benefit analysis for the decisions that really matter to you. And then throw it away.

Institutional Choice and Public Policy

To this point I've skirted around a big problem for expected value theory and cost-benefit analysis. This is the problem of how to compare the apples of cost with the oranges of benefits. For institutions—including the government—it's necessary to compare costs and benefits with the same yardstick. It would be nice if we could compare costs and benefits in terms of "human welfare units" or "utilitarian points." But no one has come up with a sensible way to calculate those things. So normally we're left with money.

Consider how one might do a cost-benefit analysis for a highly complicated policy decision. An example might be whether it pays to have high-quality prekindergarten day care for poor minority children. Such an analysis has actually been carried out by the Nobel Prize–winning economist James Heckman and his colleagues.[3] Alternative actions—high-quality day care versus no day care—are easy to specify. Heckman and company then had to identify the affected parties and estimate benefits over some period, which they arbitrarily determined would end when the children reached age forty. They had to convert all costs and benefits to monetary amounts and pick a discount rate. They didn't have to estimate the probability and value of all the cost and benefit outcomes because some of these were known from previous research; for example, savings on welfare, savings due to lowered rate of special education and retention in grade, cost of college attendance for those who went to college, and increase in earnings by age forty. Other outcomes had to be estimated. The cost of the high-quality day care compared to the cost of ordinary day care (or lack of day care at all) provided to children in the control group was estimated, though not likely to be too far off.

Heckman and company calculated the cost of crime based on the contention that crime costs $1.3 trillion per year. This in turn was based on estimates of the number and severity of crimes derived from national statistics. But the crime cost estimate is shaky. National statistics on crime, I'm sorry to tell you, are unreliable. Estimates of the number and type of crimes committed by the preschoolers by the age of forty, based on individuals' arrest records, are obviously also very uncertain. The reduction in likelihood of abuse or neglect for an individual as a child, and then later when that child becomes an adult, is difficult to assess or assign a monetary value. Heckman and company simply assign it a value of zero.

TABLE 2

Economic benefits and costs of the Perry Preschool Program calculated by Heckman (2006). All values are discounted at 3 percent and are in 2004 dollars. Earnings, Welfare, and Crime refer to monetized value of adult outcomes (higher earnings, savings in welfare, and reduced costs of crime). K–12 refers to the savings in remedial schooling. College/adult refers to tuition costs. (Reproduced by permission of *Science* magazine.)

Child care	$986
Earnings	$40,537
K–12	$9,184
College/adult	–$782
Crime	$94,065
Welfare	$355
Total benefits	$144,345
Total costs	$16,514
Net present value	$127,831
Benefit-to-cost ratio	8.74

Identifying all the parties ultimately affected by the high-quality day care seems impossible. Calculating costs and benefits for this unknown number of people therefore can't be done. And, in fact, Heckman and colleagues didn't include all the known benefits. For example, people who had been in the high-quality program were less likely to smoke, providing difficult-to-calculate benefits both for the individual in question and for untold numbers of other people, including those of us who pay higher insurance premiums because of the need to treat smoking-related diseases. Monetary costs to the victims of crime were reckoned in dollar terms only; costs for pain and suffering were apparently not calculated.

Finally, how do we assign a value to the increased self-esteem of the people who had been in the program? Or the greater satisfaction they gave to other people in their lives?

Plenty of unknowns here. But Heckman and his colleagues assigned a value to the program anyway. They calculated the benefit-to-cost ratio as 8.74. Nearly nine dollars returned for every dollar spent. This is an awfully precise figure for an analysis with so many loose ends and guesstimates. I trust that in the future you'll take such analyses by economists with a grain of salt.

But though the results of the cost-benefit analysis are a convenient fiction, was the exercise pointless? Not at all. Because we now get to the final stage of sensitivity analysis. We know that many of the numbers are dubious in the extreme. But suppose the estimate of the cost of crimes avoided is exaggerated by a factor of ten. The net benefit remains positive. More important still, Heckman and company left out many benefits either because they were not known or because it's so manifestly pointless to try to estimate their monetary value or probability.

Since there are no known significant costs other than those in Table 2, and it's only benefits that we're missing, we know that the high-quality day care program was a success and a great bargain. Moreover, the point of conducting the cost-benefit analysis was an attempt to influence public policy. And, as the saying goes, "In the policy game, some numbers beat no numbers every time."

When Ronald Reagan became president in 1981, one of his first acts was to declare, over the strong objections of many on the left, that all new regulations issued by the government should be subjected to cost-benefit analysis. The policy has been continued by all subsequent presidents. President Obama ordered that all *existing* regulations be subjected to cost-benefit analysis. The administrator responsible for carrying out the order claims that the savings to the public have already been enormous.[4]

How Much Is a Human Life Worth?

Some of the most important decisions that corporations and governments make concern actual human lives. That's a benefit (or cost) that has to be calculated in some way. But surely we wouldn't want to calculate the value of a human life?

Actually, no matter how repellent you find the concept, you're going to have to agree that we must place at least a tacit value on a human life. You would save lives if you put an ambulance on every corner. But you're not willing to do that, of course. Though the money spent on the ambulances might result in saving perhaps a life or two per week in a medium-size city, the expense would be prohibitive and you wouldn't then have the resources to provide adequate education or recreational facilities or any other public good, including (nonambulance) health care. But exactly how much education are you willing to sacrifice in order to have a reasonable number of ambulances in a city? We can be explicit or we can be tacit. But whatever decision we reach, we will have placed a value on a human life.

So what *is* the value of a human life? You may want to shop among government agencies for the answer.[5] The Food and Drug Administration valued a life in 2010, apparently arbitrarily, at $7.9 million. That represented a jump from two years previously, when the value assigned a life was $5 million. The Department of Transportation figured, also apparently arbitrarily, $6 million.

There are nonarbitrary ways of placing a value on a life. The Environmental Protection Agency values a life at $9.1 million (or rather did in 2008).[6] That's based on the amount of money people are willing to pay to avoid certain risks, and on how much extra money companies pay their workers to get them to take additional risks.[7] Another way of estimating the value of a life is to see how much we actually pay to save the life of a particular human. Economists at the Stanford Graduate School of Business made this calculation based on how much we pay for kidney dialysis.[8] Hundreds of thousands of people are alive at any one time who would be dead were it not for kidney dialysis treatments. The investigators determined that a year of "quality adjusted life" costs \$129,000 for people in dialysis, so we infer that society places a value of \$129,000 on people's quality-adjusted lives. (The quality adjustment is based on a reckoning that a year in a dialysis patient's life, which isn't all that enjoyable, is on average worth only half what an unimpaired year of life is worth. Dementia and other disabilities are more common for dialysis patients than for people of the same age who are not on dialysis.) The dialysis-based analysis puts a human life of fifty years as being worth \$12.9 million (\$129,000 × 2 × 50).

Economists call values derived in these particular nonarbitrary ways *revealed preferences*. The value of something is revealed by what people are willing to pay for it—as opposed to what they say they would pay, which can be very different. Verbal reports about preferences can be self-contradictory as well as hard to justify. Randomly selected people say they would spend about as much to save two thousand birds from suffering due to oil damage as other randomly selected people say they would spend to save two hundred thousand of the same birds.[9] Apparently people have a budget for oil-endangered birds that they won't exceed no matter how many are saved!

The great majority of developed nations have hit upon \$50,000 as the value of a quality-adjusted year of human life for the purposes of public or private insurance payment for a given medical procedure. That figure is based on no scientific determination. It just seems to be what most people consider reasonable. The \$50,000 figure means that these countries would pay for a medical procedure costing \$500,000 if it was to save the life of an otherwise healthy seventy-five-year-old with a life expectancy of ten years. But not \$600,000 (or \$500,001 for that matter). Countries would pay up to \$4 million to save the life of a five-year-old with a life expectancy of eighty-five. (The United States doesn't have an agreed-upon value of a life for purposes of insurance coverage—yet—though opinion polls show that the great majority of people are at least somewhat comfortable with calculations of that sort.)

But how about the life of someone from a less developed nation, say Bangladesh or Tanzania? Those countries are not as rich as the developed countries, but surely we wouldn't want to say that the lives of their citizens are worth less than ours.

Actually, we do say that. Intergovernmental agencies calculate that the value of a citizen of a developed country is greater than that of a citizen of a developing country. (On the other hand, this practice does have its benign aspects from the standpoint of the citizens of less developed nations. The Intergovernmental Panel on Climate Change assumes that a developed nation can pay fifteen times as much as a developing nation to avert a death due to climate change.)

I trust that by now you're dubious about techniques for calculating the value of a human life. And I haven't even started to regale you with stories such as those about the insurance companies that pay less for the life of a coal miner than for the life of an office worker on the grounds that the coal miner's value on his life is revealed to be lower because of his choice of a hazardous occupation! Or the report that the Ford Motor Company decided not to have a recall of its Pintos to put a safer gas tank in the cars because the recall would have cost the company $147 million, versus a mere $45 million for payments for wrongful deaths!

But . . . we really do have to have *some* base value for a human life. Otherwise we risk spending large amounts of money to carry out some regulation resulting in a trivial increase in the number of quality years of human life while failing to spend a modest amount of money to increase the number of quality years of human life by hundreds of thousands.

The Tragedy of the Commons

A problem for cost-benefit theory is that my benefit can be your cost. Consider the well-known *tragedy of the commons*.[10] There is a pasture that is available to everyone. Each shepherd will want to keep as many sheep in the pasture as possible. But if everyone increases the number of sheep in the pasture, at some point overgrazing occurs, risking everyone's livelihood. The problem—the tragedy—is that for each individual shepherd the gain of adding one sheep is equal to +1, but the contribution to degradation of the commons is only a fraction of −1 (minus one divided by the number of shepherds who share the pasture). My pursuit of my self-interest combined with everyone else's pursuit of their self-interest results in ruin for us all.

Enter government, either self-organized by the affected parties themselves or imposed by an external agent. The shepherds must agree

to limit the number of sheep each is allowed, or a government of some
kind must establish the limits.

Pollution creates a similar tragedy of the commons. I greatly enjoy
my plane travel, my air-conditioning, and my automobile trips. But this
makes everyone's environment more dangerous and unpleasant by increasing the pollutants in the air and ultimately by changing the climate
of the earth in potentially disastrous ways. These *negative externalities*,
as economists refer to them, harm everyone on the planet. I am hurt by
the pollution and climate change, too, of course. But my guilty pleasures have a summed value of +1 for me and the costs for me are

$$\frac{-1}{7,000,000,000}$$

Self-governing of the 7 billion of us is out of the question at the level
of the individual. "Self-government" at the level of the community of
nations is the only form possible.

The idea of cost-benefit analysis dealt with throughout this chapter
is not a novel concept to anyone. It's clear that we've been doing something like it all our lives. There are some implications of cost-benefit
theory, however, that are not at all obvious. Some of those have been
presented in this chapter. As you'll see in the next chapter, we can have
several kinds of less than optimal outcomes because of our failure
to recognize and apply some nonobvious implications of cost-benefit
theory.

Summing Up

*Microeconomists are not agreed on just how it is that people make decisions
or how they should make them.* They do agree, however, that cost-benefit
analysis of some kind is what people normally do, and should do.

*The more important and complicated the decision, the more important
it is to do such an analysis.* And the more important and complicated
the decision is, the more sensible it is to throw the analysis away once
it's done.

*Even an obviously flawed cost-benefit analysis can sometimes show in
high relief what the decision must be.* A sensitivity analysis might show

that the range of possible values for particular costs or benefits is enormous, but a particular decision could still be clearly indicated as the wisest. Nevertheless, have a salt cellar handy when an economist offers you the results of a cost-benefit analysis.

There is no fully adequate metric for costs and benefits, but it's usually necessary to compare them anyway. Unsatisfactory as it is, money is frequently the only practical metric available.

Calculations of the value of a human life are repellent and sometimes grossly misused, but they are often necessary nonetheless in order to make sensible policy decisions. Otherwise we risk spending great resources to save a few lives or fail to spend modest resources to save many lives.

Tragedies of the commons, where my gain creates negative externalities for you, typically require binding and enforceable intervention. This may be by common agreement among the affected parties or by local, national, or international agencies.

5. Spilt Milk and Free Lunch

Have you ever walked out of a restaurant without finishing a meal you had paid for because you didn't particularly like it?

Do you think an economist would say that walking out under these circumstances was a wise decision?

Suppose you are just about to go into a theater to see a play for which you had bought a ticket costing fifty dollars—and you feel that's about what the play is worth to you. Unfortunately, you have lost the ticket. Do you think you would buy another ticket for fifty dollars, thereby having paid a total of one hundred dollars to see the play?

Do you pay people to do work around your home that you don't enjoy, such as gardening or painting or cleaning?

A hospital in your town is about to be torn down to make way for a new one. It would cost about as much to renovate the old hospital, which had been extremely expensive to construct, as to build a new hospital. Would you favor renovation or construction?

You may answer such questions differently after you've read this chapter. Cost-benefit theory has some implications that are subtle but profoundly important for our daily lives. These implications are almost as important as the theory's main requirement that we choose the option with the greatest net benefit. As it happens, the implications can be derived logically from that requirement—you're likely to realize that you violate them all the time. Finding out about them is going to save you money and time. It will also improve the quality of your life.

Sunk Costs

Let's say you bought tickets a month ago to a basketball game in a city thirty miles from your home. Tonight is game night. However, the star is not playing, so the game is going to be less interesting than you thought it would be; and it's begun to snow. The tickets cost eighty dollars apiece. Do you still go to the game or do you resign yourself to staying home? WWED? What Would an Economist Do?

An economist would tell you to do a thought experiment: Suppose you hadn't bought the tickets. You intended to, but it slipped your mind. And suppose a friend called you and told you he had tickets to the game but wasn't going himself; you could have his tickets for free. If your answer would be, "Yes, that's great; I'll be right over to pick them up," then you should by all means go to the game you paid for. But if your answer would be, "You've got to be kidding. The star isn't playing and it's starting to snow," then you shouldn't go to the game even though it means your money got you nothing. If you feel uncomfortable making that decision, it's because you haven't fully incorporated the *sunk cost* principle into your decision making.

The sunk cost principle says that only future benefits and costs should figure in your choices. The money you paid for the basketball game is long gone—it's sunk—and you can't get it back by going to the game. You should go to the game only if you think your net benefit would be positive. Go if you would say to yourself, "Well, the star isn't playing and it's snowing, which is a pain. But I really feel like watching a game tonight. I've read all I care to in the newspaper and there's nothing on TV." Otherwise, don't go to the game because that would constitute paying a cost to justify a cost that can't be retrieved.

The fact that the old hospital in your town was extremely expensive to build is absolutely irrelevant to a choice of whether to renovate it or raze it and build anew. The taxes your grandparents paid to construct that hospital are a dim memory, and they're not going to reappear because you've decided to let it stand. A decision to keep or destroy that hospital must be made only with respect to the future. The net benefits for what you get in the way of a new hospital compared to the net benefits of a renovated hospital are the only considerations that count.

Should you eat a lousy meal that cost a pretty penny? Not unless you're too poor to buy peanut butter to make yourself a sandwich when you get home. You might ask for your money back if you found a fly in your soup, but you're probably not going to demand to see the manager and tell him that you're refusing to pay for the crummy lasagna. So the money for the meal is sunk. No point in incurring the additional cost of eating the darn thing.

Should you walk out of a movie that cost fifteen dollars, which you're not particularly enjoying and which shows no prospect of getting better? Absolutely.

The economist's motto, and it should be yours, is that the rest of your life begins now. Nothing that happened yesterday can be retrieved. No use crying over spilt milk.

Policy makers who are not economists often spend your money for no better reason than to rescue money they've already spent. "True, this weapon system is not very good, but we've already spent $6 billion of the taxpayers' money, and we don't want to waste that." You should remind your representatives of the adage "Don't throw good money after bad." That bad money is sunk. Even more sinister is the politician who urges continuing a war, putting more lives at risk, "so that the fallen shall not have died in vain."

Drug companies sometimes justify exorbitant prices for a drug by citing the need to "retrieve the cost of developing it." They're pulling your leg. The development money is gone. They're going to charge whatever the market will bear for the new drug—even if the development costs for the drug were very small. They get away with their claim because the public doesn't fully understand the concept of sunk costs.

A little warning, though. If you begin to live in awareness of the sunk cost principle, you're going to make the occasional mistake. I no longer walk out of plays—because I began to realize it could be demoralizing to the actors when they see that empty seat after intermission. And I no longer ask my wife whether she wants to stay to watch the rest of a movie that's boring me silly. A couple of times we had an awkward exchange: "Do you like this movie?" "Well, kind of. But we can leave if you want." "No, that's okay, I don't mind staying." And then we both sit there unhappy—my wife because she knows I'm staying in my seat even though I don't want to, and I because I've lessened her pleasure in the movie.

Speaking of spouses, some people I know, after encountering the sunk cost concept, have suggested it means people shouldn't stay in a marriage just because they've spent a long time and a lot of energy in the marriage—because that time and energy are sunk. I would be very careful about that kind of reasoning. Time and energy spent in a marriage *do* count as reasons to stay in it. If the time and energy had value previously it may have value in the future. Consider the saying "Marriage is for getting over the periods of unlove."

Opportunity Cost

It used to bother me that my mother would drive all over town to get the best bargain on detergent by cashing in two dollars' worth of coupons she had clipped from the paper. There was a hidden cost to that driving around. Money was being spent for gas and maintenance on her car. Moreover, she could have been reading a novel or playing bridge, activities that I think she valued more. In other words, she was incurring *opportunity costs* by driving around town looking for bargains.

An opportunity cost is defined as the cost of engaging in a given course of action, thereby losing the benefits of the next-best action. This principle holds where resources are limited and the chosen action precludes taking any other action. The cost is not the *sum* of the unchosen alternatives but just the *best* unchosen alternative. Anything of value can figure into opportunity costs—money, time, or pleasure.

A farmer who raises wheat forgoes the benefit of raising corn. A child who successfully tries out for the school soccer team may be forgoing the pleasure of playing football for the school or playing in the orchestra.

Life is full of opportunity costs. They can't be avoided. What *can* be avoided is paying an opportunity cost for an action that is less valuable to you than some other action you could just as easily have taken.

Economists don't mow their lawns or wash their cars. But should *you* mow your own lawn? Only if (a) you enjoy doing it or (b) you're so low on cash that you can't afford the luxury of lying in a hammock and watching your fourteen-year-old neighbor mow it. If you mow your own lawn, there are other things you can't do that you might enjoy more—working in the garden, for example, which might give you more pleasure both in the doing and in the result.

The person who drives a car rather than taking public transportation is out of pocket for the car, plus gas, plus maintenance, plus insurance—money that could have been used for travel or a housing upgrade. But the cost of owning a car tends to be hidden after it's bought, and the cost of a daily commute by bus and the occasional cab ride is quite salient. So the cost of driving a car seems slight (I've got the thing, might as well use it), whereas every trip by another means hurts a bit (fifteen bucks just to go downtown?!). As it happens, many young people have learned the principle that every car trip usually costs a lot by comparison to the alternatives. They're buying fewer cars than their parents (helped along in this by the appearance on the scene of Zipcars and their imitators).

A person who uses an office in a building the person owns is likely to consider the office to be rent-free. And an accountant might indeed record her as paying nothing for rent. But in fact she is paying something for using the office, namely the payment for the office if she were to rent it out. If the person could find an office that was as good or better than her own, but that costs less money than what the person could get for her own office, she is paying an opportunity cost for using her own office. That cost is hidden but nonetheless real.

There's a familiar slogan I find helpful in avoiding opportunity costs: "There ain't no such thing as free lunch." (The expression comes from Depression-era bars that attracted patrons by advertising free lunch. The lunch was free but the beer wasn't.) Any action you take means you can't take some other action that, upon reflection, you might prefer.

Entry-level construction and factory jobs are beginning to pay more now that home building is taking off and some manufacturing is returning to the United States. Should colleges increase student aid to attract young people who might be tempted to go for one of those jobs? An economist would point out that as salaries go up, so does the opportunity cost of going to college. If tuition at the university is $10,000 per year and the potential student could make $40,000 per year in a construction or factory job (up from the $30,000 it was a few years ago), the opportunity cost of going to college has been increased by $40,000 (assuming graduation in four years). Most economists would say it's proper for the university to respond to this opportunity cost by providing more scholarship aid to lower-income students. But I know from my own research that most academics rebel at that. "I don't want to bribe people to go to college."

It can sometimes be quite difficult to see that the value of the unchosen alternative is actually greater than that of the chosen alternative. Every hire you make for your company constitutes an opportunity cost. If there's no one more capable who can be hired, it's tempting to feel that nothing has been lost. But if there are good reasons to believe that in the near future someone more qualified could be hired, then the present hire involves an opportunity cost to the company that might indicate that hiring should be put off.

It's important to keep in mind that there are costs for being too aware of opportunity costs just as there are for being too aware of sunk costs. When I was in graduate school, I had a friend who was great fun to be with. He was always coming up with interesting things to do. If we went for a walk, after a while he might suggest that we take a bus across town to see a parade. Partway through the not terribly interesting parade, he might note that if we got a quick dinner, there would be just enough time to see a new movie we both wanted to see. After the movie, he might suggest we visit a friend who happened to live in the neighborhood.

Now, each change in activity my friend suggested, taken just by itself, was an improvement over the current activity, thereby avoiding an opportunity cost. But taken as a whole, my time with my friend was less enjoyable than it would have been without the constant calculation of new pleasures to be had. Calculation of opportunity costs can be a cost in its own right.

And back to my mother. I eventually realized that my attitude that shopping is a necessary evil to be curtailed as much as possible is not everyone's attitude. My mother would rather hunt for bargains than do most of the other things she does. Plus it's an excuse to get out of the house. So I was wrong to feel my mother was incurring net opportunity costs by her shopping.

Are the Economists Right?

How do we know that the economists are right—that we should make our choices in line with cost-benefit theory, including its sunk cost and opportunity cost corollaries?

What do the economists have to say that might convince us? They make two arguments.

1. Cost-benefit theory is logic-tight. It's based on a few assumptions that most people agree are reasonable guides for good decision making: more money is better than less, decision time counts as a cost, future benefits are of less value than present benefits, and so forth. If you agree with the assumptions, then you must buy the model because it follows mathematically from the assumptions.
2. Less common, and possibly usually made tongue-in-cheek, is the argument that cost-benefit analysis must be beneficial because corporations pay for experts to apply cost-benefit analysis to their operations. Corporations are not dumb, and they know what they want, so by implication cost-benefit rules are the correct ones to abide by.

Are you convinced by these arguments? I'm not.

Deriving appropriate behavior from a logical construction is just not very persuasive to me. An argument can be logical without being correct (see Chapter 13 on formalisms). Before we can accept an argument based on logic, we need to consider how our susceptibility to social influence and a host of other factors that operate outside of consciousness might make formal arguments less than completely convincing. And remember from the previous chapter that optimization was the normative recommendation before Herbert Simon came along and said that it's actually satisficing that's the best policy. And there's not much evidence that satisficing is what people actually do or even what they're capable of doing. So maybe they're right not to satisfice. Maybe there is another principle they're following that some theorist in the future is going to recognize as the most rational strategy, given our cognitive limitations. A good normative theory for how to make choices needs to take into consideration the Part I issues of rationality, the extent to which we are capable of self-knowledge, and the appropriate role of the unconscious in decision making. Because most psychologists believe these things, they tend to be dubious about economists' descriptions of choice behavior and their prescriptions for it.

Corporations pay for cost-benefit analysts, all right. But they also pay for handwriting analysts to assess personality, lie detector technicians, feng shui "experts," motivational speakers to hop around a stage, and astrologers. None of these is proven to be effective. Astrology has been shown to have no predictive validity whatsoever, and there's a great deal of evidence indicating zero validity of both lie detectors and handwriting analysis for any purpose a corporation might care about.

So what would convince you that you ought to use the cost-benefit principles?

What if you knew that the more familiar people are with cost-benefit principles in the abstract, the more likely they are to use them? That would be somewhat persuasive to me. As economists would be the first to insist, we must presume people to be rational until proven otherwise. If people change their behavior in order to be consistent with the abstract principles once they know them, that counts as evidence of a sort that the principles are useful.

And in fact, Richard Larrick, James Morgan, and I have found that people do use cost-benefit principles in proportion to how much they have been taught about them.[1] Economics professors are far more likely to endorse choices made on the basis of cost-benefit principles than are biologists or humanities professors. Students who have had economics courses are more likely to know the principles in the abstract and more likely to report making choices consistent with them than students who have not had economics courses (though not much more likely).

But findings such as these are contaminated by *self-selection* (see Chapter 11). People are not randomly assigned to be economists versus something else such as lawyers or bricklayers. Maybe economists are smarter than biologists, or maybe they were sympathetic to cost-benefit issues before they became economists—in fact *became* economists precisely for that very reason. And maybe students who take economics courses are smarter than students who don't and are more likely to understand and use the rules independent of how many economics courses they've taken.

Of course, for the above alternative explanations to be viable, it would have to be the case that, other things being equal, smarter people report making choices more in line with economic theory than people who are less smart. This is in fact the case. SAT and ACT verbal scores are a pretty good proxy for IQ. The correlation between SAT (and ACT) verbal score and reported use of the rules is about .4—not a huge correlation but certainly not trivial in its implications for how people should lead their lives.[2] (The correlation holds both for students who have taken economics courses and for those who haven't.)

I've conducted experiments showing that teaching cost-benefit principles to people in brief sessions—presenting even less material than you've seen in this chapter—increases the likelihood that they will endorse choices made using those principles. Even when people are tested weeks later, in the context of a telephone poll apparently unrelated to the experiment, they're more likely to endorse choices that follow from the rules.

So smarter people, and people educated in the rule system, are more likely to use the principles than less smart, untrained people. Are they better off doing so? If they're so smart, why aren't they rich?

They are richer, actually. Faculty members at the University of Michigan who report making decisions in line with cost-benefit analysis make significantly more money.[3] The relationship is even stronger for biologists and humanities professors than it is for economists (perhaps because all economists are well aware of the principles and there isn't much variation among them in that respect). And the more economics training the biologists and humanists have had, the more money they make. Moreover, I've found there is a strong tendency for raises over the past five years to be correlated with the degree to which faculty report using cost-benefit principles for their choices.

Students who report making choices in line with cost-benefit rules get better grades than students who don't. And it's not just because the rule users are smarter. In fact, the relationship between rule use and grades gets *stronger* when verbal SAT/ACT is pulled out of the equation. At every level of verbal ability, it's the students who use the rules more who get the better grades.

Why should use of cost-benefit rules make people more effective? In part because use of the rules encourages you to focus your energy where it will do the most good and drop projects that don't look like they're panning out. Avoiding the sunk cost trap and attending to opportunity costs, in other words. Some of the best advice I ever got was from a person who told me to have three categories of projects: very important and urgent, important and should be done soon, and somewhat important but no rush. Then make sure you are only working on the first category at all times, never on the other two. Not only will you be more effective, there will be more time left to goof off and enjoy yourself. (Though I do make an exception for activities with unknown payoff that might produce food for thought—especially if they're pleasurable in their own right. Henry Kissinger's advisor urged him to stop studying political science and start reading more novels.)

Summing Up

Expended resources that can't be retrieved should not be allowed to influence a decision about whether to consume something that those resources were used to obtain. Such costs are sunk, no matter what you do, so carrying out the action for which the costs were incurred makes sense only if there is still a net benefit from it. No point in eating sour grapes just because they were expensive. Corporations and politicians get the public to pay for goods and projects in order to justify past expenditures because most people don't understand the sunk cost concept.

You should avoid engaging in an activity that has lower net benefit than some other action you could take now or in the future. You shouldn't buy a thing, or attend an event, or hire a person if such an action likely blocks a more beneficial action. At least that's the case when immediate action is not strictly necessary. You should scan a decision of any consequence to see whether opportunity costs may be incurred by it. On the other hand, obsessive calculation of opportunity costs for small matters is a cost in itself. True, you can't have chocolate if you choose vanilla, but get over it.

Falling into the sunk cost trap always entails paying unnecessary opportunity costs. If you do something you don't want to do and don't have to do, you automatically are wasting an opportunity to do something better.

Attention to costs and benefits, including sunk cost and opportunity cost traps, pays. The thinkers over the centuries who have urged some form of cost-benefit analysis are probably right. There's evidence that people who make explicit cost-benefit decisions and avoid sunk costs and opportunity costs are more successful.

6. Foiling Foibles

Suppose a person needs to sell some stock to get the down payment for a house. The person owns two stocks: ABC company, which has done well recently, and XYZ corporation, which has lost money. He sells the ABC stock rather than the XYZ stock because he doesn't wish to lock in his losses on XYZ by selling it. Good idea or bad idea?

Suppose out of the kindness of my heart I give you a hundred dollars. Then I ask you to bet on a coin toss that could result in your either losing that hundred dollars or getting some larger amount of money. What is the amount that would tip you over into taking the bet? $101? $105? $110? $120? More?

The preceding chapters showed that there are many ways in which we fail to follow the precepts of cost-benefit theory. This chapter deals with several other anomalies, and it shows how we can avoid them, protecting ourselves against our tendencies to make uneconomical decisions. We don't always behave in the fully rational way demanded by cost-benefit theory, but we can arrange the world so that we don't have to in order to get the same benefits we would if we were professional economists.

Loss Aversion

We have a general tendency to avoid giving up what we already have, even in situations where the cost-benefit considerations say that we should relinquish what we have for the clear prospect of getting something better. The tendency is called *loss aversion*. Across a wide range of situations, it appears that gaining something only makes you about half as happy as losing the same thing makes you unhappy.[1]

We pay dearly for our aversion to loss. Many people would be reluctant to sell a stock that's been going down rather than a stock that's been going up. Taking a sure loss versus a possible gain is painful. People chronically sell winners, congratulating themselves on their gains, and keep losers, congratulating themselves on having avoided a certain loss. Other things being equal, a stock that's going up is more likely to keep going up than a stock that's going down is to turn around and start going up at an equal rate. Ditching winners and keeping losers over a lifetime, rather than the opposite, is the difference at retirement between being poor and very poor (or between being rich and very rich).

You can also demonstrate with gambles the extent to which we find the prospect of loss to be aversive. Suppose I ask you whether you want to make a bet. Heads you win X dollars, tails you lose $100. If X were $100, it would be a fair bet. How much does X have to be for you to be willing to take the gamble? If X were even $101, the bet would be slightly in your favor. If it were, say, $125, it would be a terrific deal. Surely worth taking unless you're so poor that the chance of loss would constitute an unacceptable risk. But a majority of people require X to be somewhere around $200, which of course is wildly in their favor. So the prospect of winning $200 is required to meet the prospect of losing $100.

Consider the following experiment, which has been carried out in dozens of business school classes. Half the students in the class are given a coffee mug with the university logo prominently displayed on it. Unlucky students who did not get a mug are asked to examine one and say how much they would pay for a mug just like it. Mug owners are asked how much they would sell their mugs for. There is a heavy discrepancy between the two amounts. On average, owners are willing to sell only when the price is double what the average nonowner is willing to pay.[2] Loss aversion lies behind this *endowment effect*. People don't want to give up things they own, even for more than they originally considered a fair price. Imagine you bought a ticket to a football game for two hundred dollars but would have been willing to pay five hundred dollars. Then a couple of weeks later you discover on the Internet that there are lots of desperate people willing to pay up to two thousand dollars for a ticket. Do you sell? Maybe not. There can be a huge spread between what something was worth to buy and what it's worth to sell—for no better reason than that we would have to give the thing up.[3]

The performing arts presenters at my university make good use of the endowment effect in their promotional campaigns. Sending people a twenty-dollar voucher they can use for ticket purchase nets 70 percent more ticket sales than mailing them a letter with a promo code for a twenty-dollar discount. People don't want to lose money by failing to cash in on the voucher they possess; but they're willing to forgo the possible gain of using the promo code when they buy their tickets.

Research by a team headed by the economist Roland Fryer found that offering to increase pay for teachers if the academic achievement of their students improved had no effect on student performance. Giving teachers the same amount of money at the beginning of the term and telling them they would have to pay back that amount if their students failed to meet a specified target resulted in a significant positive effect on student performance.[4]

It's not possible to justify the endowment effect in cost-benefit terms. I should be willing to sell a commodity at the same or slightly higher price than I paid for it. Even economists are susceptible to a range of biases, including the endowment effect bias, which prevent them from being fully rational in cost-benefit terms. The endowment effect concept, in fact, first occurred to the economist Richard Thaler when he thought about the behavior of an economist colleague who was a wine enthusiast. The man never paid more than thirty-five dollars for a bottle of wine but was sometimes unwilling to sell a bottle bought at that price even for amounts as large as one hundred dollars.[5] Having such a large spread between buying price and selling price can't be defended in terms of the normative rules of cost-benefit theory.

The previous point requires a huge qualification. Sentimental value is properly considered when thinking about a transaction. You couldn't afford to buy my wedding band. But few people have an attachment to a bottle of Château de Something-or-Other that they would describe as sentimental.

Changing the Status Quo

Loss aversion produces inertia. Changing our behavior usually involves a cost of some kind. "Shall I change the channel? I have to get up to find the remote. I have to decide what would be a more interesting program to watch. Or maybe I would enjoy reading a book more. What book? Oh, well, I haven't watched *Jeopardy!* in quite a while. Might be fun."

TV networks are well aware of this sort of sluggishness in our behavior and schedule their most popular programs early in prime time, with the expectation that many watchers will stay tuned to their channel after the popular program is over.

The biggest problem with loss aversion is that it prompts a *status quo bias.*[6] I continue to receive several newsletters that I long ago stopped reading because the time is never right to figure out how to stop the darn things from coming. Right now I'm in the middle of X (watering the garden, making a list of what to buy at the hardware store, getting organized to write a paper). Canceling the newsletters means ceasing to do something that I value. So I'll do it tomorrow when I have nothing much else to do. (Ha!)

The economist Richard Thaler and the legal scholar Cass Sunstein have shown numerous ways we can make the status quo bias work in our favor.[7] Some of the most important work rests on a single concept, namely "default option."

Only 12 percent of Germans allow the government to harvest their organs, but 99 percent of Austrians do. Who would have thought the Austrians were so much more humanitarian than the Germans?

Actually, there's no reason to assume there's a difference between Germany and Austria in concern for their fellow citizens. Austrians have an *opt-out* policy for organ harvest. The default presumption is that the organs of dead people are available for transplanting. You have to tell the state you *don't* want to donate. Germany has an *opt-in* policy. The default is that the state has no right to harvest a person's organs unless the person specifically agrees to that. Opt-in is the policy in the United States. Scores of thousands of people have died who would have lived if the United States had an opt-out policy.

Choice architecture plays a vital role in determining what decisions people make. Some ways of structuring decisions result in better outcomes for individuals and for society than other ways of structuring decisions. No one is hurt by opt-out procedures for things like organ donation; no coercion is involved because people who wish not to have their organs harvested are free to decline. The deliberate design of decision frameworks that function for individual and collective good has been called "libertarian paternalism" by Thaler and Sunstein.[8]

The difference between choice architectures that foster the right choices and those that don't can be subtle—at least to people who are unfamiliar with the power of loss aversion and consequent status quo bias.

In a "defined contribution" retirement plan, an employer pays a fixed amount of money into a savings plan equal to some fraction of what the employee puts into the plan. For example, an employer might match the employee's contribution up to 6 percent of the employee's salary. Both employer and employee contributions are invested, and the money is available at retirement. The nature of the investment—individual stocks or bonds or mutual funds—is determined by the employee. The benefit is unknown—that depends on how the investments do. Employees are offered the defined contribution plan upon being hired. The plan is portable, unlike "defined benefit" plans, such as those automobile companies and many state and local governments offer, where one knows in advance how much one will receive at a given age.

One would think that virtually everyone would take advantage of the free money offered by employers who provide defined contribution plans. In fact, however, about 30 percent of employees fail to sign up for such plans.[9] A study in Britain of twenty-five corporations that offered defined contribution plans—and paid 100 percent of the cost—found that barely half of employees signed up for the plan![10] This is like burning up a portion of your salary.

A sensible choice architecture for savings plans would not require people to opt in, which after all takes little more effort than checking a box, but would have an opt-out default, which requires even less energy than that. You are enrolled in the plan unless you ask not to be. In one plan, the opt-in approach resulted in scarcely more than 20 percent enrollment three months after starting the job and only 65 percent after three years on the job. Automatic enrollment resulted in 90 percent enrollment after a few months and 98 percent after three years on the job.[11]

Even if people can be steered into enrolling in a retirement plan, this is no guarantee they'll have enough money to retire with. Typically, the amount of money people decide to put into a retirement savings plan at the beginning of employment is not enough for them to live on in retirement. How to get people to save enough?

Shlomo Benartzi and Richard Thaler have invented the Save More Tomorrow plan to deal with this problem.[12] An employee who starts out with 3 percent being saved would be told after a while on the job that higher savings are needed to have enough for retirement, and might be told that an additional 5 percent is needed immediately, with subsequent increases in future years. If the employee balks, the counselor suggests increasing the savings rate whenever there is a raise. If the raise is 4 percent, there would be an automatic increase in retirement savings of some fixed amount, such as 3 percent. This would continue until an adequate amount is being withdrawn for savings, say 15 percent. This works beautifully because it allows inertia to work in employees' favor and protects against loss aversion by ensuring that the increase in savings is not experienced as a loss.

Choice: Less Can Be More

A colleague from Germany joined my department a number of years ago and asked why Americans seemed to find it necessary to have a choice of fifty breakfast cereals. I had no answer except to say that I guess people—or Americans anyway—like to have a lot of choices.

Certainly the Coca-Cola company believes Americans like lots of choices. Which do you prefer: Coca-Cola, Caffeine-free Coca-Cola, Caffeine-free Diet Coke, cherry Coke, Coca-Cola Zero, Vanilla Coke, Vanilla Coke Zero, Diet cherry Coke, Diet Coke, Diet Coke with Lime, or Diet Coke with Stevia (in a green can!)? Or perhaps you'd rather just have a Dr Pepper.

Coke is not alone in assuming that the sky's the limit when it comes to choice. There's an upscale grocery store in Menlo Park, California, that offers 75 types of olive oil, 250 varieties of mustard, and 300 types of jams.

But are more choices always better than fewer? You would be hard-pressed to find an economist who would tell you that fewer choices are better. But it's becoming clear that more choices are not always desirable—either for the purveyor of goods or for the consumer.

The social psychologists Sheena Iyengar and Mark Lepper set up a booth at that Menlo Park grocery store where they displayed a variety of jams.[13] Half the time during the day there were six jams on the table and half the time there were twenty-four jams. People who stopped at the booth were given a coupon good for one dollar off any jam they purchased in the store. Many more people stopped at the booth when there were twenty-four jams than when there were only six. But ten times as many people bought a jar of jam when there were only six at the table than when there were twenty-four! Retailers beware: customers do sometimes recognize opportunity costs of endless examination of alternatives and buzz off when you overload them with choices.

In 2000, the Swedish government reformed its pension plan. In a move similar to George W. Bush's attempt to privatize a portion of Social Security pay-ins, the government set up an investment scheme for individuals. The plan they came up with seems sensible on the surface—to financial experts.[14]

1. Participants were allowed to choose to invest in as many as five mutual funds approved by the government for their portfolios.
2. There were 456 of these funds, each of which was allowed to advertise.
3. Exhaustive information about each of these funds was provided to participants in book form.
4. One fund, which was not allowed to advertise, was chosen by government economists to be the default fund.
5. People were encouraged to choose the funds they would invest in.

Two-thirds of the participants did in fact choose their own funds rather than accept the default. But the choosers didn't do a very good job of picking their funds. First of all, whereas the default fund charged a .17 percent management fee, the average fund selected by participants charged .77 percent—a discrepancy that over time makes a considerable difference. Second, whereas the default fund invested 82 percent in equities, the average percent chosen by other participants was 96. Sweden's economy is 1 percent of the world's economy, but the default fund chose to invest 17 percent of its equities in Swedish corporations. That's a lot of eggs to put in one small basket. But the other participants ended up with *48 percent* Swedish stocks. The default fund had 10 percent fixed-income securities, the others an average of 4 percent. The default fund had 4 percent in both hedge funds and private equities. The others put nothing into those types of investments. Finally, technology stocks had been soaring in the period just prior to the pension plan rollout. A great many investors put most or all of their investments in a fund consisting solely of ill-fated technology stocks. That fund had been up by 534 percent over the preceding five years, but if you recall the ill-fated year 2000, those stocks were about to fall over a cliff.

An economist would say that each of the differences between the default fund and the average of the others was in favor of the default fund. A psychologist would say that the deviations between the default and the others were mostly explicable in terms of a number of understandable biases.

1. I've heard of the Swedish Widget Company, but not of the American Whatsit Corporation.
2. I want (all) my money to be in the type of fund with the greatest growth potential, namely stocks.
3. Only a chump would choose a stock fund that hadn't made a great deal of money in the recent past over a fund that had been booming.
4. I don't know what the heck a hedge fund is or what a private equity is.
5. I'll read the book about the investment funds as soon as I get some time.

No economist would pick such a lopsided investment strategy as the average Swedish participant did.

But how did the funds do? It's not completely reasonable to render a judgment about quality of investment decisions on the basis of the initial seven years of performance, but in fact the default fund made 21.5 percent, versus 5.1 percent for the average of the other funds.

How should the Swedish procedures have been changed? And what should the United States do in the event that partial privatization of Social Security pay-ins eventually occurs?

The basic problem with the Swedish plan is that the government was so in thrall to the goal of choice. Many of the options on the list of funds would have been chosen by no seasoned investor. People should not have been given their choice of funds without some guidance. The government should have told people that they ought to consult with a financial expert before choosing or that they should probably just go with the default. But this is an age in which people fear being too directive.

The medical profession, by the way, is much too enamored of the choice mantra for my taste. Doctors who lay out a number of choices of treatment options for you, telling you the costs and benefits of each, but then fail to make a recommendation are not doing their job as well as they should. They have expertise that they should share with you in the form of a recommendation, or at least a default choice together with suggestions about why the other options might be things you would want to consider. My personal default as a patient: "What would you do, Doctor?"

Incentivize, Incentivize

I recently participated in a World Economic Forum panel on decision making. The panel consisted of economists, psychologists, political scientists, physicians, and policy experts. The panel's charge was to discuss ways to get people to behave in their own interest and in society's interest. The buzzword was "incentivize," and it was clear that most members of the group could think of incentives only in terms of promise of monetary gain or threat of monetary loss. Give people subsidies for wise behavior and threaten them with fines for unwise behavior.

Of course there's no question that monetary incentives can be highly effective—indeed, sometimes they can be astonishingly effective. So much so that the members of the panel were quite prepared to believe the claim that some cities have had marked success in getting teenage girls to avoid pregnancy by paying them as little as a dollar a day.[15] The program sounds like a great deal because the amount is trivial for the city but allegedly is sufficient to greatly reduce pregnancies—and consequent costs to the city, not to mention costs to the girls. But in fact, whether the program has any effect is controversial, and any success it might have could be due to other aspects of the program, such as sex education and opening up life possibilities by bringing the girls to a college campus on a regular basis. Our faith in monetary incentives leaves us too willing to believe the "dollar a day" claim.

One of the main messages of this book is that behavior is governed by a host of factors other than monetary ones, and some nonmonetary incentives are highly effective when monetary incentives are useless or worse. Social influence can do much more to move people in the desired direction than promises of reward or threats of punishment or any amount of admonition.

Mere information about the behavior of others can motivate people to change their own behavior. If I know that other people are behaving better than I am inclined to do, that serves as an agent of social influence. I want to do what others do.

Knowledge that others are behaving better than one would be inclined to think they are is often much more effective than preaching—which can backfire by suggesting that bad practices are *more* widespread than they actually are. That gets the conformity ball rolling against you.

Want to get people to use less electricity? If they're using more electricity than their neighbors, leave a hang tag on their door telling them so.[16] For good measure, add a frowny-face emoticon. And give them suggestions about how they can save energy. If they're using less energy than their neighbors, leave a hang tag on their door telling them that, too—but be extra sure to add a smiley emoticon, or the information may result in their actually increasing energy usage. So far, this clever intervention by social psychologists has resulted in savings of more than $300 million in energy costs for the state of California and prevented billions of pounds of CO_2 from escaping into the atmosphere.

Want students at your local college to engage in less binge drinking? Recall from Chapter 2 that this can be achieved by telling students how much other students on their campus are drinking, which is likely to be less than students tend to think.[17] Want to increase compliance with state tax laws? Tell people about the rate of compliance in your state. Most people greatly overestimate the amount of tax cheating that goes on in their state. When they do overestimate, they can justify their little fibs: "I'm not one of those crooks, I'm just putting some body English on my travel expenses." Information about rates of cheating makes it harder to engage in that kind of rationalization.

Want to get people to conserve water and protect the environment by reusing towels in hotel rooms? You could just ask them to do that, but it's not as effective as telling them that most guests at the hotel do reuse towels, which in turn is less effective than telling them that a majority of people who had "stayed in this room" in the past have reused towels.[18]

You can tell people that insulating their attic will save them several hundred dollars per year, and you can further promise them a monetary reward for insulating. But you're not likely to get much compliance. If you're like me, there's a big barrier: the attic is so full of junk that it's hard to reach the ceiling to put in the insulation. Try offering a subsidy to help people move their junk around or to just get rid of it, and see whether that doesn't increase attic insulation.

Monetary incentives and attempts at coercion are particularly likely to be counterproductive if the person perceives the incentive or coercion threat as indicating that the activity is not a very attractive one. Why else offer an incentive to perform the activity or issue a threat if they don't perform it?

Many years ago Mark Lepper, David Greene, and I put an intriguing new activity out on the table of a nursery school.[19] Children could draw with a type of felt-tip markers they hadn't encountered before. We observed the children and recorded the amount of time each spent drawing with the markers. Two weeks later, an experimenter approached some of the children and asked them whether they would like to draw some pictures for him using the markers in order to have a chance to win a Good Player Award: "See? It's got a big gold star and a bright blue ribbon, and there's a place here for your name and your school. Would you like to win one of these Good Player Awards?" Other children were simply asked whether they would like to draw with the markers. All children who "contracted" to draw with the markers were given the Good Player Award. Some children did not "contract" for drawing with the markers, but the experimenter gave them one anyway. And some did not contract for the award and did not get one. One to two weeks later, the marker activity was again placed on the table.

Children who got the award after having contracted to draw with the markers in order to win it drew with the markers less than half as much as children who got an unanticipated award or no reward at all. The young contractors realized that drawing with the markers was something they did in order to get something they wanted. The other children could only infer that they were drawing with the markers because they wanted to.

As Mark Twain said, "Work consists of whatever a body is *obliged* to do, and . . . Play consists of whatever a body is not obliged to do."

We should all aspire to think like an economist steeped in cost-benefit principles. But that's a tall order (even for economists). Fortunately, this chapter shows there's a lot we can do to arrange our lives, and those of people we care about, to make an end run around our failings.

Summing Up

Loss considerations tend to loom too large relative to gain considerations. Loss aversion causes us to miss out on a lot of good deals. If you can afford a modest loss in order to have an equal chance for a larger gain, that's the way you should normally bet.

We're overly susceptible to the endowment effect—valuing a thing more than we should simply because it's ours. If you have an opportunity to divest something at a profit but feel reluctant to do so, ask yourself whether that's simply because of your ownership of the thing rather than some other factor such as expected net value for keeping the thing. Sell your white elephants no matter how much room you have in your attic for them. The people who tell you to give away every article of clothing you haven't used for a year are right. (Do what I say, not what I do. I periodically shuffle shirts around in my closet that I haven't worn in a decade because there is after all a chance I might buy a jacket that one of them would look good with.)

We're a lazy species: we hang on to the status quo for no other reason than that it's the way things are. Put laziness to work by organizing your life and that of others so that the easy way out is actually the most desirable option. If option A is better than option B, give people option A as the default and make them check a box to get option B.

Choice is way overrated. Too many choices can confuse and make decisions worse—or prevent needed decisions from being made. Offer your customers A or B or C. Not A through Z. They'll be happier and you'll be richer. Offering people a choice implies that any of the alternatives would be rational to pick; spare people the freedom of making a wrong choice in ignorance of your opinion of what would be the best alternative. Tell them why you think option A is best and what the considerations are that might make it rational to choose something different.

When we try to influence the behavior of others, we're too ready to think in terms of conventional incentives—carrots and sticks. Monetary gain and loss are the big favorites among incentives. But there are often alternative ways of getting people to do what we want. These can be simultaneously more effective and cheaper. (And attempts at bribery or coercion are remarkably likely to be counterproductive.) Just letting people know what other people do can be remarkably effective. Want people to use less electricity? Tell them they're using more than their neighbors. Want students to drink less alcohol? Tell them their fellow students are not the lushes they think they are. Rather than pushing people or pulling people, try removing barriers and creating channels that make the most sensible behavior the easiest option.

PART III

CODING, COUNTING, CORRELATION, AND CAUSALITY

I have been speaking prose all my life, and didn't even know it.
—Monsieur Jourdain, in Molière's *The Bourgeois Gentleman*

Like Molière's bourgeois gentleman, who was delighted to discover he had been speaking prose all his life, you may be surprised and pleased to discover that you've been making statistical inferences all your life. The goal of the next two chapters is to help you to make better statistical inferences and more of them.

Regardless of whether you think you know how to do statistics, you need to read these chapters.

This is true if either of the following is the case.

a) *You don't know much statistics*. If that's true, these chapters are the most painless way you'll ever find to gain sufficient knowledge to be able to use statistics in everyday life. And you simply can't live an optimal life in today's world without basic knowledge of statistics.

You may feel that statistics is too boring or difficult for you to trudge through. My sympathies. When I was in college, I was desperate to become a psychologist, and that was going to be impossible unless I took a statistics course. But I had little math background and I was scared witless for the first few weeks of what I thought was a course in mathematics. But eventually I realized that the math in basic inferential statistics doesn't go much beyond knowledge of how to extract a square root. (These days the knowledge that's required for that is to be aware of the location on your calculator of the square root button.) Some theorists believe statistics isn't a branch of mathematics at all but rather a set of empirical generalizations about the world.

To relax you even more, I can tell you that all the statistical principles explained here—and they're the ones that are most valuable for everyday life—are commonsensical. Or at least on a little reflection they satisfy common sense. You already know how to apply most of the principles in at least some circumstances, so many if not most of the shocks you'll get in these chapters will be shocks of recognition.

b) *You know a fair amount about statistics, or even a lot.* If you quickly peruse the statistical terms in the next two chapters, you may feel that you have little to learn from them. I assure you that is not the case. Statistics is normally taught in order to prevent if at all possible its use in any domain except IQ tests and agricultural yields. But statistical competence will escape to an unlimited number of everyday life domains if you learn how to frame events in such a way that statistical principles are immediately relevant.

Psychology graduate students at most universities take two or more statistics courses during their first two years. Darrin Lehman, Richard Lempert, and I tested students on their ability to apply statistical principles to everyday life problems, and on their ability to critique scientific claims, at the beginning of their graduate careers and again two years later.[1] Some students gain hugely in their ability to apply these principles to everyday life and some gain little.

The students who gain ability to apply statistics to everyday life events tend to be those in the so-called soft areas of psychology—social psychology, developmental psychology, and personality psychology. The low gainers are those in the hard areas of psychology—biopsychology, cognitive science, and neuroscience.

Since they've all taken the same statistics courses, why do the soft-area students learn more than the hard-area students? It's because the soft-area students are constantly applying the statistics they've learned to everyday life kinds of events. Which behaviors of mothers are most associated with social confidence in infants? How do we code and measure mothers' behaviors and how do we assess and measure social confidence? Do people change their evaluations of objects simply by virtue of being given the objects? How do we measure their evaluation of objects? How much more talking in small groups is done by extroverts compared to introverts? How should we code amount of talking: Percent of time each person talks? Number of words? Should we count interruptions separately?

In short, the soft-area students learn to do two things that this chapter will help you to do: (1) *frame* everyday life events in such a way that the relevance of statistical principles is obvious and you can make contact with them, and (2) *code* the events in such a way that approximate versions of statistical rules can be applied to them. The next two chapters do that with anecdotes and realistic problems that can crop up in everyday life. The chapters are intended to help you build *statistical heuristics*—rules of thumb that will suggest correct answers for an indefinitely large number of everyday life events. These heuristics will shrink the range of events to which you will apply only intuitive heuristics, such as the representativeness and availability heuristics. Such heuristics invade the space of events for which only statistical heuristics are appropriate.

Two years of thinking about rats or brains or memory for nonsense syllables produces little improvement in ability to apply statistical principles to everyday life events. Students in the hard areas of psychology may learn scarcely more than students in chemistry and law. I found that students in those fields gain literally nothing over two years in the way of ability to apply statistics to the everyday world.

I also studied medical students, expecting that they would gain very little in ability to think statistically about everyday life problems. I was wrong. The students improved a fair amount. I attended the University of Michigan's medical school for a few days to find out what might account for the improvement. To my surprise, the medical school does require some training in statistics, in the form of a pamphlet that is handed out early on. Probably much more important than the rather minimal formal training in statistics, students learn about medical conditions and human behavior in potentially quantifiable ways and reason about them in explicitly statistical terms. "The patient has symptoms A, B, and C and does not have D and E. What is the likelihood that the patient has Disease Y? Disease Z? Disease Z, you say? You're probably wrong about that. Disease Z is quite rare. If you hear hoofbeats, think horses, not zebras. What tests would you want to order? Tests Q and R, you say? You're wrong. Those tests are not very statistically reliable; moreover they're quite expensive. You might order test M or N, which are cheap and statistically reliable, but neither is a very valid predictor of either disease Y or disease Z."

Once you have the knack of framing real-world problems as statistical ones and coding their elements in such a way that statistical heuristics can be applied, those principles seem to pop up magically to help you solve a given problem—often without your conscious awareness that you're applying a rough-and-ready version of a statistical principle.

I'll introduce in ordinary language some basic statistical principles that have been around for one hundred years or more. Scientists in many fields use these concepts to determine how confident they can be that they've characterized an object in the right way, to estimate the strength of relationships between events of various kinds, and to try to determine whether those relationships are causal. As we'll see, they can also be used to illuminate everyday problems and help us make better decisions at work and at home.

7. Odds and *N*s

In 2007, Governor Rick Perry of Texas issued an executive order mandating that all twelve-year-old girls in Texas receive a vaccination for human papilloma virus, which can result in cervical cancer. In her effort to score points on Rick Perry in the Republican primary of 2012, the candidate Michele Bachmann announced that a woman had told her that her "little daughter took that vaccine, that injection, and she suffered from mental retardation thereafter."

What was wrong with Bachmann's inference—or at least her invitation to us to infer—that HPV vaccinations cause mental retardation? Let us count the ways.

We need to think of Bachmann's evidence as a report about a *sample* of the *population* of all twelve-year-old girls in the United States who got the vaccination. One case of mental retardation constitutes a very small sample (low *N*) and isn't remotely enough to establish that the population of girls receiving the injection was at risk from it.

In fact, there have been several rigorous *randomized control studies* with follow-ups in which girls were randomly selected either to be inoculated or not. These studies all had a very large *N*—number of cases. In none of those studies were vaccinated girls found to have a higher rate of retardation than girls who were not vaccinated.

Bachmann's sample of twelve-year-old girls who had the vaccination consisted of one case—an example of reliance on the *Man Who* statistic, as in "I know a man who." Bachmann's sample was *haphazard* at best rather than *random*. The closer that sample-selection procedures approach the gold standard of random selection—for which the definition is that every individual in the population has an equal chance of appearing in the sample—the more we should trust them. If we don't know whether a sample is random, any statistical measure we conduct may be *biased* in some unknown way.

Actually, Bachmann's sample is not even as good as a haphazard one. Assuming Bachmann was telling the truth, she had a strong motive to present this one case to the public. And she may not have been telling the truth, or her informant may not have been telling the truth. Which is not to say that the informant was lying. She may well have believed what she allegedly told Bachmann. If her daughter had the vaccination and then was diagnosed with mental retardation, it's possible that the mother's reasoning is an instance of the *post hoc ergo propter hoc* error: after this, therefore because of this. The fact that thing 1 precedes thing 2 doesn't mean it necessarily caused thing 2. In any case, it seems to me that we should regard Bachmann's claim as not quite reaching the very low bar set by the Man Who statistic.

One of my favorite examples of the post hoc ergo propter hoc error combined with the Man Who statistic was provided to me by a friend who overheard the following conversation between two elderly men. First man: "My doctor told me I have to quit smoking or I'll die from it." Second man: "No! Don't quit! I have two friends who quit smoking because their doctors told them to and they were both dead within a few months."

Sample and Population

Recall the hospital problem from Chapter 1 on inferences. Smaller hospitals will have more days than larger hospitals when the number of baby boys is greater than 60 percent of the total births. The only way to know that is to understand the *law of large numbers*: sample values such as means and proportions come closer to the actual values in the population the larger the N, that is, the larger the sample.

At the extremes of population size, it's easy to see the force of the law of large numbers. Suppose there were ten births at a given hospital on a particular day. How likely would it be that 60 percent or more of the births were male? Quite a good chance, of course. We wouldn't be suspicious of a coin that came up heads six times out of ten. Suppose there were two hundred births at another hospital on a particular day. How likely would that deviant value be? Extremely unlikely, obviously. That's exactly comparable to tossing an allegedly fair coin two hundred times and coming up with 120 or more heads instead of the expected one hundred.

As an aside, I'll note that the accuracy of a sample statistic (mean, median, standard deviation, etc.) is essentially independent of the size of the population the sample is drawn from. Most national polls in elections sample about a thousand people, for which they claim accuracy of about + or –3 percent. A sample of a thousand people gives about as good an estimate of the exact percentage supporting a given candidate when the population size is 100 million as when the population size is ten thousand. So when your candidate is ahead by eight points, don't be concerned when the other candidate's campaign manager pooh-poohs the poll, saying millions of people will vote and the poll sampled only a thousand people. Unless the people who were polled are not typical of the population in some important respect, the campaign manager's candidate is well on the way to being toast. Which brings us to the issue of *sample bias*.

The law of large numbers holds only for samples that are *unbiased*. A sample is biased if the procedure for obtaining it allows for the possibility that a given sample value is in error. If you're trying to find out what proportion of the workers in a factory would like to have flexible work time and you sample only men, or only people who work in the cafeteria, those people could be different in some important respect from the factory population as a whole, giving an erroneous estimate of the proportion of workers who favor flex time. If there's bias in the sample, the larger the sample, the more confident we become about a wrong estimate.

It has to be noted that the national polls don't actually sample randomly from the population. In order for that to be the case, every voter in the country would have to have an equal chance of being in the sample. When this is not the case, you run the risk of serious bias. One of the very first national election polls in the United States, conducted by the now defunct *Literary Digest*, reported that Franklin Roosevelt was about to lose the 1932 election, which he in fact won in a landslide. The *Digest*'s problem? Its poll was taken by telephone—and only well-off (and disproportionately Republican) people had telephones at the time.

A similar source of bias characterized some of the polls in the 2012 election. The Rasmussen polling firm did not call cell phone lines, neglecting the fact that people who have only cell phones are disproportionately likely to be young and Democrat-leaning. The Rasmussen poll systematically overestimated support for Romney compared to polls that sampled both landline numbers and cell phone numbers.

Once upon a time, when people answered their phones and opened their doors to interviewers, you could come close to a random sample of the population. These days poll accuracy depends in part on the pollster's data and intuitions about how to doctor the sample—weighting it by stirring into the brew of poll numbers the likelihood that respondents will vote, respondents' party identification, gender, or age, how the community or region has voted in the past, eye of newt, toe of frog . . .

Finding the True Score

Consider the following pair of problems.

University X has a renowned musical theater program. The program awards scholarships to a small number of high school graduates who show unusual promise. Jane, the director of the musical theater program, has friends who are drama teachers at high schools in the area. One afternoon she goes to Springfield High to watch a student who has been highly touted by her teachers as a superb young actress. She sees a rehearsal for a Rodgers and Hammerstein musical in which the student has the lead role. The student flubs several lines, seems not to have the correct conception of the character she's playing, and comes across as having little stage presence. The director tells her colleagues that she now doubts the judgments of her teacher friends at the high school. Is this a wise conclusion or not?

Joe is a talent scout for the football team at university Y. He visits high school practice sessions all over the state, looking for prospects whose coaches have told Joe he ought to consider them for his team. One afternoon he goes to Springfield High to see a quarterback who has an excellent win-loss record, has impressive TD and completion-percentage statistics, and is highly praised by his coaches. At the practice, the quarterback misthrows several passes, is sacked a few times, and gains little yardage overall. The scout's report states that the quarterback has been overrated and recommends that the university cease to consider him for recruitment. Is this a wise recommendation or not?

If you said that Jane is wise and Joe is not, it's a fair bet that you're knowledgeable about sports but not dramatics. If you said Joe is wise and Jane is not, it's a fair bet that you're knowledgeable about dramatics but not sports.

I find that people who don't know much about sports often say Joe is probably right that the quarterback may not be all that talented, but people who do know sports are likely to think Joe may be being too hasty. They recognize the possibility that Joe's (rather small) sample of the quarterback's behavior may be an extreme performance and that there is a strong possibility that the quarterback's ability is closer to Joe's informants' evaluation than to his own.

People who don't know much about dramatics are likely to say that the actress probably isn't very good, but people who do know dramatics think Jane may be being too dismissive of the judgments of Jane's friends at the high school. Other things being equal, the more you know about a given domain, the more likely you are to be able to use statistical concepts to think about that domain. In this case, the important concept is the law of large numbers.

Here's why the law of large numbers is relevant. A quarterback's performance over one or more seasons can be assumed to be a pretty reliable indicator of his skill. If his coaches back that up with the insistence that he is really excellent, we have a great deal of evidence—lots of data points—indicating that the quarterback that Joe observes really is extremely good. Joe's evidence is pretty piddling by comparison: one set of observations on one day.

The inherent variability of a player's performance, even the inherent variability of the performance of an entire team, is recognized in the adage that on any given Sunday, any team in the NFL can beat any other team in the NFL. This certainly doesn't mean that all teams are equally talented, it just means that you need a fairly large sample of behavior to determine with confidence the relative skill of different teams.

The same logic applies to the theater program director's opinion of the actress she sees. If several people who know the actress well say that she's highly talented, then the director ought to place relatively little weight on the sample she has. I find that very few people recognize this, except for people who have done some acting and are well aware of the variability of performance in that field. In his autobiography, the comedian and actor Steve Martin writes that almost any comedian can be great some of the time. The successful ones are those who can be at least good all the time.

In statistical lingo, the coach and the theater program director are trying to determine the *true score* of the candidate they are observing. The observation = true score + error. This is the case for measurements of all kinds, even the height of people or the temperature of the atmosphere. There are two ways to improve the accuracy of the score. One is to get better kinds of observations—better yardsticks or thermometers. The other way is to "cancel out" whatever errors there are in your measurements by obtaining a larger number of observations and averaging them. The law of large numbers applies: the more observations you make, the closer you get to the true score.

The Interview Illusion

Even when we're highly knowledgeable about some domain, and highly knowledgeable about statistics, we're likely to forget the concept of variability and the relevance of the law of large numbers. The Department of Psychology at the University of Michigan interviews its top applicants for graduate study before it makes a final decision as to whether to admit a student. My colleagues tend to put substantial weight on their twenty- to thirty-minute interviews with each candidate. "I don't think she's a good bet. She didn't seem to be very engaged with the issues we were discussing." "He looks like a pretty sure thing to me. He told me about his excellent honors thesis and made it clear he really understands how to do research."

The problem here is that judgments about a person based on small samples of behavior are being allowed to weigh significantly against the balance of a much larger amount of evidence, including college grade point average, which summarizes behavior over four years in thirty or more academic courses; Graduate Record Exam (GRE) scores, which are in part a reflection of how much a person has learned over twelve years of schooling and in part a reflection of general intellectual ability; and letters of recommendation, which typically are based on many hours of contact with the student. In fact, college grade point average has been shown to predict performance in graduate school to a significant degree[1] (a correlation of about .3, which as you'll see in the next chapter is rather modest), and GRE scores also predict performance to about the same degree. And the two are somewhat independent of each other, so using them both improves prediction above the level of each separately. Using letters of recommendation adds yet a bit more to the accuracy of predictions.

But predictions based on the half-hour interview have been shown to correlate less than .10 with performance ratings of undergraduate and graduate students, as well as with performance ratings for army officers, businesspeople, medical students, Peace Corps volunteers, and every other category of people that has ever been examined. That's a pretty pathetic degree of prediction—not much better than a coin toss. It wouldn't be so bad if people gave the interview as much weight as it deserves, which is little more than to let it be a tiebreaker, but people characteristically undermine the accuracy of their predictions by overweighting the value of the interview relative to the value of other, more substantial information.

In fact, people overweight the value of the interview so much that they're likely to get things backward. They think an interview is a better guide to academic performance in college than is high school GPA, and they think an interview is a better guide to quality of performance in the Peace Corps than letters of recommendation based on many hours of observation of the candidate.[2]

To bring home the lesson of the interview data: Given a case where there is significant, presumably valuable, information about candidates for school or a job that can be obtained by looking at the folder, *you are better off not interviewing candidates*. If you could weight the interview as little as it deserves, that wouldn't be true, but it's almost impossible not to overweight it because we tend to be so unjustifiably confident that our observations give us very good information about a person's abilities and traits.

It's as if we regard the impression we have of someone we've interviewed as resulting from an examination of a hologram of the person—a little smaller and fuzzier to be sure, but nevertheless a representation of the whole person. We ought to be thinking about the interview as a very small, fragmentary, and quite possibly biased sample of all the information that exists about the person. Think of the blind men and the elephant, and try to force yourself to believe you're one of those blind men.

Note that the interview illusion and the fundamental attribution error (FAE) are cut from the same cloth, and both are amplified by our failure to pay sufficient attention to the quantity of evidence that we have about a person. A better comprehension of the FAE, namely that we overestimate the relevance of stable dispositions relative to situations, would lead us to be dubious about how much we can learn from an interview. A firmer grasp of the law of large numbers makes us less vulnerable to both the FAE and the interview illusion.

I wish I could say that my knowledge of the utility of interviews always makes me skeptical about the validity of my own conclusions based on interviews. My understanding of that principle has a limited dampening effect, however. The illusion that I have valuable and trustworthy knowledge is too powerful. I just have to remind myself not to weight the interview—or any brief exposure to a person—very heavily. This is especially important when I have presumably solid information about the person based on other people's opinions formed after long acquaintance with the candidate, as well as records of academic or job achievements.

I have no difficulty, however, in remembering the limitations of your judgments based on a brief interview!

Dispersion and Regression

I have a friend—let's call her Catherine—who consults with hospitals about management practices. She loves her job in part because she likes to travel and meet new people. She's something of a gourmet and enjoys going to restaurants that she has reason to believe are very good ones. But she says that she's typically disappointed when she goes back to restaurants that she initially thought were excellent. Second meals rarely seem as good to her as the first. Why do you suppose this is?

If you said "maybe the chefs change a lot" or "maybe her expectations are so high that she's likely to be disappointed," you're ignoring some important statistical considerations.

A statistical approach to the problem would start by noting that, after all, there is a chance element to how good a meal Catherine gets at any given restaurant on any given occasion. For any individual sampling a given restaurant on different occasions, or for a group of diners who eat at a given restaurant at a given time, there will be variation in judged meal quality. That first meal Catherine gets at the restaurant could be anywhere from so-so (or even worse) to fabulous. This variation is what makes us refer to judged meal quality as a *variable*.

Any kind of variable that is *continuous* (there's a complete range of measurements from one extreme to the other, for example in the case of height), as opposed to discontinuous (for example in the case of gender or political affiliation) is going to have a *mean* and a *distribution* around the mean. Given that fact alone, it's not surprising that Catherine is often disappointed: it's a virtual certainty that some of the time her second experience of a restaurant is going to be worse than the first (as well as some occasions when the second meal is better than the first).

But we can say more than that. We can *expect* that Catherine's opinion of a restaurant where she's had one excellent meal is going to go down. This is because of the fact that the closer to the mean a given value is, the more common it is. The farther from the mean, the rarer the value. So if she's had an excellent meal on occasion 1, the next meal is likely to be less extreme. This is true for all variables that meet the definition of *normal distribution*, which is captured by the so-called *bell curve*, shown in Figure 2.

The normal distribution is a mathematical abstraction, but it's approximated surprisingly often for continuous variables—the number of eggs laid weekly by different hens, the number of errors per week in the manufacture of car transmissions, and the IQ test scores for people are all arrayed in something approximating a normal distribution. No one knows why that's the case; it just is.

There are a number of ways of describing the *dispersion* of cases around the mean. One is the *range*—the highest value in the available cases minus the lowest value. A more useful measure of dispersion is the *average deviation* from the mean. If the average quality of meals that Catherine assigns to the first meal in each of the restaurants she eats in in different cities is, say, "pretty good," and the average deviation from that mean is equal to, say, "very good" on the plus side, and just "fairly good" on the minus side, we would say that the degree of dispersion—the average deviation—around Catherine's mean of first-meal-quality judgments is not very great. If the average deviation ranged from "superb" on the plus side to "rather mediocre" on the minus side, we'd say that the dispersion was quite large.

But there's a much more useful measure of dispersion that we can calculate for any variable that can be given continuous numerical values. This is the *standard deviation*. (Or SD, the symbol for which is the Greek sigma: σ.) The standard deviation is (essentially) the square root of the average of the squared distance of each observation from the mean. Conceptually it's not all that different from the average deviation, but the standard deviation has some extremely useful properties.

The normal curve in Figure 2 is marked out into standard deviations. About 68 percent of values are within plus and minus one standard deviation from the mean. As an example, consider IQ test scores. Most IQ tests are scored so that the mean is arbitrarily set at 100 and the standard deviation at 15. Someone with an IQ of 115 is a standard deviation above the mean. The distance between the mean and one standard deviation above the mean is pretty large. Someone with an IQ of 115 could be expected to finish college and might even do some postgraduate work. Typical occupations would be professional, managerial, and technical. Someone with an IQ of 100 would be more likely to have some community or junior college study or just high school work and to have an occupation such as store manager, clerk, or tradesman.

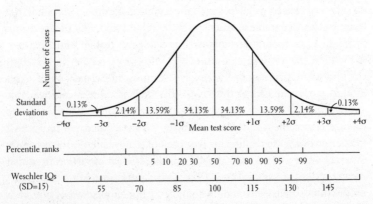

Figure 2. Distribution of IQ scores around the mean of 100, with corresponding standard deviations and percentile ranks.

Another set of useful facts about the standard deviation concerns the relation between percentiles and standard deviations. About 83 percent of all observations are less than one standard deviation (SD) above the mean. An observation exactly at one SD from the mean is at the eighty-fourth percentile of the distribution. Sixteen percent of the remaining observations are above the eighty-fourth percentile. Almost 98 percent of all observations lie below two SDs above the mean. A score exactly at two SDs from the mean is at the 98th percentile. Just over 2 percent of the remaining observations are above that. Nearly all observations will fall between three SDs below the mean and three SDs above it.

Knowing the relation between standard deviations and percentages is helpful for judgments about most of the continuous variables we encounter. For example, the standard deviation is a measure frequently used in finance. The standard deviation of the rate of return on an investment is a measure of the volatility of the investment. If a given stock has averaged a 4 percent rate of return for the past ten years, with a standard deviation of 3 percent, this means that your best guess is that 68 percent of the time in the future the rate of return will be between 1 percent and 7 percent, and 96 percent of the time it will be greater than −2 and less than 10 percent. That's pretty stable. It may not make you rich, but it's probably not going to send you to the poorhouse. If the standard deviation is 8, that means that 68 percent of the time the rate of return is going to be between −4 percent and +12 percent. You could do very well indeed with the stock. Sixteen percent of the time you'll get more than a +12 percent return. On the other hand, 16 percent of the time you'll lose more than 4 percent. That's fairly volatile. And 2 percent of the time you will lose more than 12 percent; 2 percent of the time you will gain more than 20 percent. You could make a killing or lose your shirt.

So-called value stocks have low volatility both with respect to dividends and with respect to price. They might pay out 2, 3, or 4 percent every year and they're probably not going to go up very much in a bull market or down very much in a bear market. So-called growth stocks generally have returns with a larger standard deviation, meaning much greater upside potential paired with significantly greater downside risk.

Financial advisors generally counsel young clients to go for growth and hold on through both bull and bear markets, because over the long haul growth stocks do indeed tend to grow—though the dips can be unnerving. Financial advisors usually counsel older clients to switch to mostly value stocks so they don't get caught in a bear market just as they are about to retire.

Interestingly, all of what you've just read about normal distributions holds independent of the shape of the normal distribution, which only sometimes resembles a bell curve. Curves can be *kurtotic* (bulging) in various ways. *Leptokurtic* (slender) curves look like a rocket ship in a 1930s comic book and have very high peaks and short tails. *Platykurtic* (broad) curves look like a boa constrictor that swallowed an elephant and have low peaks and long tails. Nevertheless, for both distributions, 68 percent of all values lie within plus and minus one standard deviation.

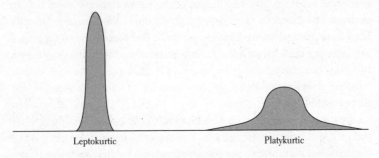

Leptokurtic   Platykurtic

But back to our question of why it is that Catherine is typically disappointed when she returns to a restaurant where she got an excellent meal. We've agreed that her evaluation of restaurant meals is a variable: it ranges from, say, execrable (1st percentile) to, say, ambrosial (99th percentile). Let's say that an excellent meal is one that is about at Catherine's 95th percentile or higher—better than about 94 percent of the meals she eats. Now ask yourself the following question about your own meal experiences. Do you think it's more likely that every meal you might happen to eat in a restaurant where you haven't eaten before would be excellent, or that only some meals would be excellent? If you think that you wouldn't expect all meals to be excellent, and you happen to get an excellent one the first time, then the expected value of a second meal is at least slightly lower than the quality of that excellent first meal.

Catherine's second-meal experiences can be thought of as instances of *regression to the mean*. If meal experiences are distributed normally, extreme values are by definition unlikely, so an event of a given kind following an extreme event of that kind is likely to be less extreme. Extreme events regress to less extreme ones.

Regression effects are lying in plain sight everywhere. Why is the baseball rookie of the year so often a disappointment the second year? Regression. The rookie's performance the first year was an outlier from his true score and he has no way to go but down. Why is the stock that grew in value more than any other in year 1 so often a mediocre performer or worse in year 2? Regression. Why does the worst-performing kid in the third grade perform a little better the next year? Regression. None of this is to say that the only thing going on is regression. It's not that the mean of a distribution is a black hole that sucks all extreme observations into it. Any number of other things could be operating to push performance level up or down. But in the absence of knowing exactly what those things are, we have to recognize that extreme scores are usually followed by less extreme ones because the combination of forces resulting in extreme values is not likely to sustain itself over time and trials. The rookie of the year happened to have a coach who was having an unusually good year; he played against relatively weak opponents in his first few games, which built his confidence; he just got engaged to the girl of his dreams; his health was perfect; he had no injuries that impaired his play, and so on. The next year an injured elbow kept him out of several games; the coach moved to another team; there was a serious illness in his family. Or whatever. And there will always be whatevers.

Two questions for which the regression principle is (surprisingly) relevant: (1) What is the likelihood that an American aged twenty-five to sixty will have an income placing the person in the top 1 percent in a given year? (2) What is the likelihood that the person will be in the top 1 percent for ten consecutive years?

The chances that a person will have a top 1 percent income once are over 110 in 1,000. Bet you wouldn't have guessed that. The chances that a person will do that ten years in a row are 6 in 1,000. Very surprising given the odds for one year. These figures seem surprising because we don't spontaneously think about a value like income as being highly variable and thus susceptible to substantial regression effects. But there is in fact a great deal of variability for an individual's income from year to year (especially at the high end of the income distribution). Extreme incomes are surprisingly common in the population as a whole. But precisely because they are extreme, such incomes are not likely to be repeated all that frequently. The great majority of those much-resented 1 percenters are on their way down, so you may want to take it easy on them!

The same kinds of figures apply for low incomes. More than 50 percent of Americans will be impoverished or nearly so at least once in their lives; conversely, not all that many people are in poverty forever. People perennially on the dole are a rarity. The great majority of people who are ever on welfare are there only for a couple of years or so.[3] You may want to take it a little easier on people on welfare, too.

We can make some pretty serious mistakes by failing to conceptualize events in terms of the possibility of regression to the mean. The psychologist Daniel Kahneman once told a group of Israeli flight instructors that praise was more effective than criticism at changing someone's behavior in the direction that's desired.[4] One of the instructors contradicted him, saying that praise of a pilot's execution of a maneuver seemed to make the pilot worse at that maneuver, whereas yelling at the pilot for a poor execution resulted in improvement the next time around. But the instructor had failed to pay sufficient attention to the fact that novice pilot performance is a variable, and regression to the mean is expected after an especially good performance—or an especially bad performance. An execution that's better than average is expected on probabilistic grounds alone to be followed by something closer to the mean, that is, worse. An execution worse than average is expected to get better.

In all probability the flight instructor was getting worse performance from his students than he could have achieved if he had conceptualized performance as a continuous variable for which you have to expect that any extreme value is going to be followed by a less extreme value. He would then be giving positive reinforcement to better-than-average performance, making him a better teacher.

The flight instructor's error gets a boost from a two-edged cognitive sword we all carry around. We are superb causal-hypothesis generators. Given an effect, we are rarely at a loss for an explanation. Seeing a difference in observations over time, we readily come up with a causal interpretation. Much of the time, no causality at all is going on—just random variation. The compulsion to explain is particularly strong when we habitually see that one event typically occurs in conjunction with another event. Seeing such a correlation almost automatically provokes a causal explanation. It's tremendously useful to be on our toes looking for causal relationships that explain our world. But there are two problems: (1) The explanations come too easily. If we recognized how facile our causal hypotheses were, we'd place less confidence in them. (2) Much of the time, no causal interpretation at all is appropriate and wouldn't even be made if we had a better understanding of randomness.

Let's try a couple of other applications of the regression principle.

If the mother of a child has an IQ of 140 and the father has an IQ of 120, what is your best guess about what the IQ of the child would be?
    160 155 150 145 140 135 130 125 120 115 110 105 100

Psychotherapists speak of a hello/goodbye effect for many patients. Patients report that their condition is worse than it really is before beginning therapy and report that their condition is better than it is at the end of therapy. Why might this be?

If you said the expected value for the child's IQ—given that one parent's IQ is 140 and the other's is 120—is 140 or higher, you haven't taken into consideration the phenomenon of regression to the mean. An IQ of 120 is higher than the mean and an IQ of 140 is higher still. Unless you think the correlation between parents' IQ and child's IQ is perfect, you have to predict that the child's IQ is lower than the average of the two parents' IQs. Since the correlation between the average of two parents' IQs and that of the child is .50 (which I didn't expect you to know), the expected value of the child's IQ is halfway between the midparent value and the mean of the population, namely 115. Supersmart parents are going to have kids who are merely smart on average. And supersmart kids are going to have parents who are merely smart on average. Regression works in both directions.

The usual explanation of the hello/goodbye phenomenon is that patients fake bad in order to be eligible for therapy but want to ingratiate themselves with the therapist at the end of therapy. Regardless of whether there is some truth to this explanation, we expect patients to be better at the end of therapy than at the beginning because they're probably less emotionally healthy than usual at the time they seek therapy, and because the mere passage of time makes regression to the mean likely. You would expect the hello/goodbye effect in the absence of any treatment at all. In fact, doctors of all kinds generally have time on their side: the expected health of a patient will tend to improve over time no matter what, unless the disease is progressive. For that matter, any intervention at all has a good shot at being deemed effective. "I had some dandelion soup and my cold cleared right up." "My wife took agave root extract as soon as she got the flu and she had it for half as long as I did." The combination of the Man Who statistic plus the post hoc ergo propter hoc heuristic has made rich many a manufacturer of nostrums. And they can truthfully claim that the great majority of people got better after taking their remedies.

But in talking about regression I'm getting a little ahead of myself. The discussion has shaded from the law of large numbers to the concept of *covariation* or *correlation*. That's the topic of the next chapter.

Summing Up

Observations of objects or events should often be thought of as samples of a population. Meal quality at a given restaurant on a given occasion, performances of a given athlete in a particular game, how rainy it was during the week we spent in London, how nice the person we met at the party seems to be—these all have to be regarded as samples from a population. And all assessments that are pertinent to such variables are subject to error of some degree or other. The larger the sample, other things being equal, the more the errors will cancel one another out and bring us closer to the true score of the population. The law of large numbers applies to events that are hard to attach a number to just as much as to events that can readily be coded.

The fundamental attribution error is primarily due to our tendency to ignore situational factors, but this is compounded by our failure to recognize that a brief exposure to a person constitutes a small sample of a person's behavior. The two errors lie behind the interview illusion—our drastic overconfidence that we know what a person is like given what the person said or did in a thirty-minute encounter.

Increasing sample size reduces error only if the sample is unbiased. The best way to ensure this is to give every object, event, or person in the population an equal chance of appearing in the sample. At the very least we have to be attentive to the possibility of sample bias: Was I relaxed and in pleasant company when I was with Jane at Chez Pierre or was I uptight because my judgmental sister-in-law was also there? Larger samples just make us more confident about our erroneous population estimates if there is bias.

The standard deviation is a handy measure of the dispersion of a continuous variable around the mean. The larger the standard deviation for a given type of observation, the less confident we can be that a particular observation will be close to the mean of the population of observations. A big standard deviation for a type of investment means greater uncertainty about its value in the future.

If we know that an observation of a particular kind of variable comes from the extreme end of the distribution of that variable, then it's likely that additional observations are going to be less extreme. The student who gets the highest grade on the last exam is probably going to do very well indeed on the next exam, but isn't likely to be the one who gets the highest grade. The ten stocks with the highest performance in a given industry last year are not likely to constitute the top ten this year. Extreme scores on any dimension are extreme because the stars aligned themselves just right (or just wrong). Those stars are probably not going to be in the same position next time around.

8. Linked Up

Statistics can be helpful, and sometimes essential, in order to character-ize something accurately. Statistics are equally valuable for determining whether there's a relationship between one thing and another. As you might guess, being sure about whether a relationship exists or not can be even more problematic than characterizing a given thing accurately.

You have to characterize things of type 1 correctly as well as things of type 2. Then you have to count how frequently type 1 things occur with type 2 things, how frequently type 1 things do not occur with type 2 things, and so on. If the variables are continuous, the job gets still harder. We have to figure out whether greater values for type 1 things are associated with greater values for type 2 things. When stated in this abstract way, it seems pretty clear that we're going to have big problems in estimating the degree of association between variables. And in fact, our problems with detecting covariation (or correlation) are really very severe. And the consequences of being off base with our estimate can be very serious.

Correlation

Have a look at Table 3 below. Is symptom X associated with disease A? Put another way, is symptom X diagnostic of disease A?

The way to read Table 3 is to note that twenty of the people who have disease A have symptom X and eighty of the people who have dis-ease A don't have symptom X; ten of the people who don't have the disease have the symptom and forty of them don't have it. On the face of it, this would seem to be the simplest covariation detection task you could present to people. The data are dichotomous (either/or). You don't

TABLE 3. THE ASSOCIATION BETWEEN
DISEASE A AND SYMPTOM X

		Disease A	
		Yes	No
Symptom X	Present	20	10
	Absent	80	40

have to collect information, or code the data points and assign numerical values to them, or remember anything about the data. You don't have any prior beliefs that might make you predisposed to see one pattern versus another; and the data are set up for you in summary form. How do people perform on this very basic covariation detection task?

Pretty badly, actually.

A particularly common failing is to rely exclusively on the "Present/ Yes" cell of the table. "Yes, the symptom is associated with the disease. Some of the people with symptom X have the disease." This tendency is an example of the *confirmation bias*—a tendency to look for evidence that would confirm a hypothesis and failing to look for evidence that might disconfirm the hypothesis.

Other people who look at the table pay attention only to two cells. Some of these conclude that the symptom is associated with the disease "because more people who have the disease have the symptom than do people who do not have the disease." Others conclude that the symptom is not associated with the disease "because more people with the disease don't have the symptom than do have it."

Without having been exposed to some statistics, very few people understand that you have to pay attention to all four cells in order to be able to answer the simple question about association.

You have to compute the ratio comparing the number of people who have the disease and also have the symptom with the number of people who have the disease and don't have the symptom. You then compute the ratio comparing the number of people who don't have the disease but do have the symptom with the number of people who don't have the disease and don't have the symptom. Since the two ratios are the same, you know that the symptom is not associated with the disease any more than it is associated with not having the disease.

You might be alarmed to know that most people, including doctors and nurses whose daily lives are concerned with treatment of disease, usually fail to get the right answer when examining tables like Table 3.[1] For example, you can show them a table indicating how many people with a disease got better with a particular treatment and how many didn't and how many people with the disease who didn't get the treatment got better and how many didn't. Doctors will sometimes assume that a particular treatment helps people because more people with the treatment got better than didn't. Without knowing the ratio of the untreated who got better to the untreated who didn't, no conclusion whatsoever is possible. Tables like these, incidentally, are sometimes called the *2 × 2 table* and sometimes called the *fourfold table*.

There's a neat little statistic called *chi square* that examines the probability that the two proportions differ enough for us to be confident that there is a genuine relationship. We say that the relationship is real if the difference between the two proportions is *statistically significant*.

A typical criterion for saying that an association is significant or not is whether the test (chi square or any other statistical test) shows that the degree of association could happen by chance only five in one hundred times. If so, we say it's significant at the .05 level. Significance tests can be applied not only to dichotomous (either/or) but also to continuous data.

When the variables are continuous and we want to know how closely they're associated with one another, we apply the statistical technique of *correlation*. Two variables that are obviously correlated are height and weight. Not perfectly correlated of course, because we can think of many examples of short people who are relatively heavy and tall people who are relatively light.

A variety of different statistical procedures can tell us just how close the association between two variables is. A frequently used technique for examining the degree of association of continuous variables is one called the Pearson product moment correlation. A zero correlation means there is no association at all between two variables. A correlation of +1 means there is a perfect positive association between two

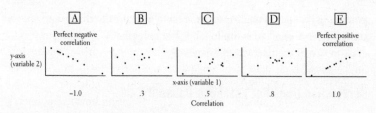

Figure 3. Scatterplots and correlations.

variables: as values on variable 1 go up, values on variable 2 go up to an exactly corresponding degree. A correlation of –1 means there is a perfect negative association.

Figure 3 shows visually, on so-called scatterplots, how strong a correlation of a given magnitude is. The individual graphs are called scatterplots because they show the degree of scatter away from a straight-line, perfect relationship.

A correlation of .3 is barely detectable visually, but it can be very important practically. A correlation of .3 corresponds to the predictability of income from IQ,[2] and of graduate school performance from college grades.[3] The same degree of predictability holds for the extent to which incipient cardiovascular illness is predicted by the degree to which an individual is underweight, average, or overweight.

A correlation of .3 is no joke: it means that if someone is at the 84th percentile (one SD above the mean) on variable A, the person would be expected to be at the 63rd percentile (.3 SD above the mean) on variable B. That's a lot better predictability for variable B than you have when you don't know anything about variable A. In that case you have to guess the 50th percentile for everybody—the mean of the distribution of variable B. That could easily be the difference between having your business thrive or go belly-up.

A correlation of .5 corresponds to the degree of association between IQ and performance on the average job. (The correlation is higher for demanding jobs and lower for jobs that are not very demanding.)

A correlation of .7 corresponds to the association between height and weight—substantial but still not perfect. A correlation of .8 corresponds to the degree of association you find between scores on the math portion of the Scholastic Aptitude Test (SAT) at one testing and scores on that test a year later—quite high but still plenty of room for difference between the two scores on average.

Correlation Does Not Establish Causality

Correlation coefficients are one step in assessing causal relations. If there is no correlation between variable A and variable B, there (probably) is no causal relation between A and B. (An exception would be when there is a third variable C that masks the correlation between A and B when there is in fact a causal relation between A and B.) If there is a correlation between variable A and variable B, this doesn't establish that variation in A *causes* variation in B. It might be that A causes B or B causes A, and the association could also be due to the fact that both A and B are associated with some third variable C, and there is no causal connection between A and B at all.

Pretty much everyone with a high school education recognizes the truth of these assertions—in the abstract. But often a given correlation is so consistent with plausible ideas about causation that we tacitly accept that the correlation establishes that there is a causal relation. We are so good at generating causal hypotheses that we do so almost automatically. Causal inferences are often irresistible. If I tell you that people who eat more chocolate have more acne, it's hard to resist the assumption that something about chocolate causes acne. (It doesn't, so far as is known.) If I tell you that couples who make elaborate wedding preparations have longer-lasting marriages, it's natural to start wondering just what it is about elaborate weddings that makes for longer marriages. In fact, a recent article in a distinguished newspaper reported on the correlation and then went on to speculate about why serious work on planning weddings would make the marriage last longer. But if you think about the correlation long enough, you'll realize that elaborate wedding preparations aren't a random event; rather, they're obviously going to be more likely for people with more friends, more time to be together, more money, and goodness knows what else. Any of those things, or more likely all of them, could be operating to make marriages more lasting. To pull one fact out of that tangled web and start to speculate on its causal role makes little sense.

Consider the associations in Box 1, all of which are real. You'll see that for some the implied causal link seems highly plausible and for others the implied link is highly implausible. Whether you think the implied causal link is plausible or not, see whether you can come up with explanations of the following types: (1) A causes B, (2) B causes A, or (3) something correlated with both A and B is causal and there is no causal link at all between A and B. See some possible answers in Box 2.

Box 1. Thinking About Correlations:
What Causal Relationships Could Be Going On?

1. *Time* magazine reported that attempts by parents to control the portions their children eat will cause the children to become overweight. If the parents of overweight children stop trying to control their portions, will the children get thinner?
2. Countries with higher average IQs have higher average wealth measured as per capita gross domestic product (GDP). Does being smarter make a country richer?
3. People who attend church have lower mortality rates than those who don't.[4] Does this mean that belief in God makes people live longer?
4. People who have dogs are less likely to be depressed. If you give a dog to a depressed person, will the person get happier?
5. States with abstinence-only sex education have higher homicide rates. Does abstinence-only sex education cause aggression? If you give more informative sex education to students in those states, will the homicide rate go down?
6. Intelligent men have better sperm—more sperm and more mobile sperm.[5] Does this indicate that attending college, which makes people smarter, also improves sperm quality?
7. People who smoke marijuana are subsequently more likely to use cocaine than people who don't smoke marijuana. Does marijuana use cause cocaine use?
8. Ice cream consumption and polio were almost perfectly correlated in the 1950s, when polio was a serious threat. Would it have been a good public health move to outlaw ice cream?

Box 2. Possible Answers to Questions About Correlations in Box 1

1. It could be that parents try to control the portions that children eat if they're overweight. If so, the direction of causation runs opposite from *Time* magazine's hypothesis. You don't make the child obese by trying to control portions; you try to control portions if the child is obese. It could also be the case that less happy, more stressful families have more controlling parents and their children are more likely to be overweight, but there's no causal connection between food policing behavior on the part of the parent and weight of the child.

2. It could be that richer countries have better education systems and hence produce people who get higher IQ scores. In that case, it's wealth that causes intelligence rather than the other way around. It's also possible that some third factor, such as physical health, influences both variables. (All three of these causal relations are real, incidentally.)

3. It could be that healthier people engage in more social activities of all kinds, including going to church. If so, the direction of causation runs opposite to the one implied: One reason people go to church is that they're healthy, but going to church doesn't make them any healthier. Or it could be that an interest in social activities such as going to church causes people both to participate in more social activities and to be healthier.

4. It could be that people who are depressed are less likely to do anything fun, such as buying a pet. If so, the direction of causation is opposite to the one implied: depression makes you less likely to get a pet. (But in fact, giving a pet to a depressed person does improve the person's mood, so pets can indeed be good for your mental health; it's just that the correlation between the two doesn't prove that.)

5. It could be that states that are poorer are more likely to have higher homicide rates and states that are poorer are more likely to have abstinence-only sex education. Indeed, both are true. So there may be no causal connection at all between sex education and homicide. Rather, poverty or low educational levels or something associated with them may be causally linked to both.

6. It could be that greater physical health helps people to be smarter and helps sperm to be of better quality. Or some other factor could

be associated with both intelligence and sperm quality, such as drug or alcohol use. So there might be no causal connection between intelligence and sperm quality.
7. It could be that people who take any kind of drug are more sensation seeking than other people and therefore engage in many kinds of stimulating behavior that are against the law. Smoking marijuana may not cause cocaine use, and cocaine use may not cause marijuana use. Rather, the third factor of sensation seeking may influence both.
8. Ice cream consumption and polio were correlated highly in the 1950s because polio is easily communicated in swimming pools. And both ice cream and swimming become more common as the weather gets warmer.

Illusory Correlation

I can't stress enough how important it is to actually collect data in a systematic way and then carry out calculations in order to determine how strong the association is between two variables. Just living in the world and noticing things can leave you with a hopelessly wrong view about the association between two events. *Illusory correlation* is a real risk.

If you think it's plausible that there's a positive relation between two variables (the more A, the more B), your casual observations are likely to convince you that you're right. This is often the case not only when there is in fact no positive correlation between the variables but even when there is actually a *negative* correlation. Noticing and remembering the cases that support your hypothesis more than ones that don't is another aspect of confirmation bias.

Conversely, if a link is implausible, you're not likely to see it even if the link is fairly strong. Psychologists have placed pigeons in an apparatus with a food pellet dispenser and a disk on the floor that can be lit up. The pellet dispenser will deliver a pellet if the disk is lit and the pigeon does *not* peck at it. If the pigeon does peck at it, there will be no food pellet. A pigeon will starve to death before it discovers that not pecking at a lighted disk will result in its getting food. Pigeons haven't made it this far by finding it plausible that not pecking something is likely to result in getting food.

People can find it as hard as pigeons to overcome presuppositions.

Experimenters have presented clinical psychologists with a series of Rorschach inkblot responses allegedly made by different patients, with the patients' symptoms printed along with the patients' alleged responses.[6] One card might show a patient who (a) saw genitals in the inkblot and (b) had problems with sexual adjustment. After perusing the set, psychologists are quite likely to report that patients who see genitals are likely to have problems with sexual adjustment, even when the data are rigged to indicate that such patients are *less* likely to have problems with sexual adjustment. It's just too plausible that sexual adjustment problems might be associated with being hypervigilant about genitals, and the positive instances stand out.

When you tell the psychologists that they're mistaken and that the series shows a *negative* association between seeing genitals and having problems with sexual adjustment—that patients who see genitals are actually less likely to have problems with sexual adjustment—the psychologists may scoff and tell you that in their clinical experience it *is* the case that people with sexual adjustment problems are particularly likely to see genitals in Rorschach blots. No, it isn't. When you actually collect the data you find no such association.

In fact, virtually no response to any Rorschach card tells you anything at all about a person.[7] Hundreds of thousands of hours and scores of millions of dollars were spent using the test before anyone bothered to see whether there was any actual association between responses and symptoms. And then for decades after the lack of association was established, the illusion of correlation kept the test in circulation, and more time and money were wasted.

I don't mean to pick on psychologists and psychiatrists with these examples. Undergraduates make exactly the same errors that clinicians do in experiments on illusory correlation using the Rorschach, reporting that seeing genitals goes with sexual problems, seeing funny-looking eyes goes with paranoia, seeing a weapon goes with hostility.

These findings can be summarized by saying that if a person (or other organism) is *prepared* to see a given relationship, that relationship is likely to be seen even when it's not in the data.[8] If you're *counterprepared* to see a given relationship, you're likely to fail to see it even when it's there. Cats will learn to pull a string to get out of a box; they will not learn that licking themselves will get them out of a box. Dogs can readily learn to go to the right to get food rather than to the left if a speaker sounds a tone on the right; only with great difficulty will dogs learn which direction to go when a higher-pitched tone indicates food is on the right and a lower tone indicates food is on the left. It just seems more likely that spatial cues are related to spatial events than it is that pitch cues are related to spatial events.

Our old friend the representativeness heuristic generates infinite numbers of prepared relationships. Genitals are representative of anything having to do with sex. Eyes are representative of suspicion. Weapons are representative of hostility. The availability heuristic also does a good job of creating prepared relationships. Films and cartoons show people with funny-looking eyes (squinting, rolling, etc.) when they're suspicious.

What if a person is neither prepared nor counterprepared to see a relationship?

What would happen if, for example, a person listened to a bunch of people say the first letter of their names and then sing a musical note—and was then asked whether there was a relationship between the position of the letter in the alphabet and the duration of the musical note?

How high does the correlation between such arbitrarily paired events have to be before people can reliably detect it?

The answer is that the correlation has to be about .6—slightly higher than the .5 correlation shown in Figure 3.9 And that's when the data come to the person all at once and the person is doing his level best to see what the relationship is. As a practical matter, this finding means that you can't rely on your belief that there's a correlation between two variables unless the association is quite strong—higher than many of the correlations on which we base the choices in our daily lives. You've got to be systematic to get it right: observe, record, and calculate or you're just blowing smoke.

An Exception

There's one important exception to the rule that covariation is very difficult to detect accurately. When two events—even arbitrary ones—occur close together in time, the covariation will usually be noticed. If you switch on a light just before delivering an electric shock to a rat, the rat will quickly learn the association between the light and the shock. But even for this sort of highly dramatic pairing of events, there is a sharp decline in ability to learn as a function of the time interval between two events. Animals—and humans—don't learn associations between arbitrarily paired events if you go much beyond a couple of minutes.

Reliability and Validity

Many years ago, a friend of mine and his wife had been trying to have a baby. After several years without success, they finally went to a fertility specialist. The news was not good. My friend's sperm count was "too low to result in impregnation by normal means." My friend asked the physician how reliable the test was. "Oh, it's very reliable," said the physician. What he meant was: the test doesn't make mistakes—it gives you the true score. He was using the term "reliable" in its lay sense of accuracy.

Reliability is the degree to which measurement of a particular variable gives the same value across occasions, or the degree to which one type of measure of a variable gives the same value as another type of measure of that variable.

Measures of height have a reliability (correlation across occasions) of virtually 1. IQ measured across occasions separated by a couple of weeks is around .9. IQ as measured by two different tests typically indicates reliability of more than .8. Two different dentists will agree about extent of decay in a tooth with a reliability of less than .8.[10] This means that not all that infrequently your tooth gets filled by Dentist Smith whereas Dentist Jones would have let it be. For that matter, any given dentist's judgments don't correlate perfectly with her own judgments on different occasions. Dr. Jones may fill a tooth on Friday that she would have left undrilled on Tuesday.

How about the reliability of sperm counts? Reliability for any given type of test for sperm count is low,[11] and reliability as indicated by the degree to which you get the same result with different measures is also low. Different ways of measuring sperm count at the same time can come up with quite different results.[12]

Validity is typically also measured by correlations. The validity of a measure is the degree to which it measures what it's supposed to measure. IQ tests have substantial validity—around .5—as measured by the degree to which IQ scores correlate with GPA in grade school. (In fact, it was the desirability of predicting school performance that motivated the early twentieth-century French psychologist Alfred Binet to create the first IQ test.)

Please note the extremely important principle that *there can be no validity without reliability*. If a given person's judgment about a variable is utterly inconsistent (for example, a correlation of zero between the person's judgments about the level of variable A on one occasion and the level of variable A on another occasion), that person's judgments can have no validity, that is, they can't predict the level of another variable B with any accuracy at all.

If test X and test Y that are supposed to measure a given variable don't agree beyond a chance level, then at most only one of those tests can have any validity. Conversely, there can be very high reliability with no validity whatsoever. Two people can agree perfectly on the degree of extroversion characteristic of each of their friends, and yet neither observer may be able to accurately predict the degree of extroversion exhibited by his friends in any given situation (as judged by objective measures of extroversion such as talkativeness or assessments by psychological experts).

Handwriting analysts claim to be able to measure honesty, hardworkingness, ambition, optimism, and a host of other attributes. To be sure, any two handwriting analysts may agree with each other quite well (high reliability), but they're not going to be able to predict any actual behavior related to personality (no validity). (Handwriting analysis can be quite useful for some purposes, though; for example, for medical diagnosis of a number of central nervous system maladies.)

Coding Is the Key to Thinking Statistically

I'm going to ask you some questions concerning your beliefs about what you think the correlation is between a number of pairs of variables. The way I'll do that is to ask you how likely it is that A would be greater than B on one occasion given that A was greater than B on another occasion. Your answers in probability terms can be converted to correlation coefficients by a mathematical formula.

Note that if you say "50 percent" for a question below, you're saying that you think there's no relationship between behavior on one occasion and behavior on another. If you say "90 percent," you're saying that there is an extremely strong relationship between behavior on one occasion and behavior on another. For the first question below about spelling ability, if you think that there is no consistency between spelling performance on one occasion and spelling performance on another occasion, you would say "50 percent." If you think that there is an extremely strong relationship between spelling performance on one occasion and spelling performance on another spelling test, you might say "90 percent." Commit yourself: write down your answer for each of the questions below or at least say your answer out loud.

1. If Carlos gets a higher grade on a spelling test than Craig at the end of the first month of fourth grade, what is the likelihood that Carlos will get a higher grade on a spelling test at the end of the third month?

2. If Julia scores more points than Jennifer in the first twenty games of the basketball season, what is the likelihood that she will score more points in the second twenty games?

3. If Bill seems friendlier than Bob on the first occasion you encounter him, what is the likelihood that he will seem friendlier on the second occasion?

4. If Barb behaves more honestly than Beth in the first twenty situations in which you observe them (paying a fair share of the bill, cheating or not while playing a board game, telling the truth about a grade in a class, etc.), what is the likelihood that Barb will behave more honestly than Beth in the second twenty situations in which you observe them?

Table 4 presents the correlations corresponding to percentage estimates of the kind you just made.

It so happens that I know the answers to these questions based on studies that have been conducted.[13] I know the correlation between performance on one spelling test and performance on another and between the average of twenty spelling tests and the average of another twenty spelling tests, between how friendly a person seems on one occasion and how friendly a person seems on another occasion and between friendli-

TABLE 4. THE CONVERSION OF PERCENTAGE ESTIMATES INTO CORRELATION COEFFICIENTS

Percentage Estimate	Correlation	Percentage Estimate	Correlation
50	0	75	.71
55	.16	80	.81
60	.31	85	.89
65	.45	90	.95
70	.59	95	.99

ness averaged over twenty situations and then over another twenty situations, and so on.

I'm betting that your answers showed the following pattern.

1. Your answers indicate that you think the correlation between basketball performance in twenty games and performance in another twenty games is high, and higher than the correlation between scores on one spelling test and scores on another.
2. Your answers indicate that you think that the correlation between friendliness on one occasion and friendliness on another occasion is quite high, and about as high as the correlation between honesty on twenty occasions and honesty on another twenty occasions.
3. Your answers indicate that the correlations for traits are higher than the correlation for abilities.

At any rate, that describes the guesses of the college student participants in the experiment that I did with Ziva Kunda.[14]

Take a look at Figure 4. Note that people's guesses about behaviors that reflect abilities (averaging over the actual data for spelling and basketball) are close to the facts. The correlation between behavior (spelling or points scored in basketball) in one situation and another is moderately

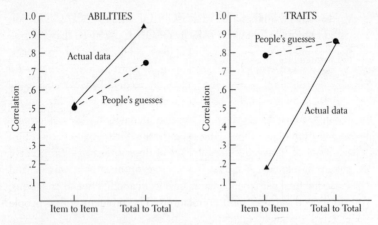

Figure 4. People's guesses about correlations based on small and large amounts of data for abilities (averaged over spelling and basketball) and for traits (averaged over friendliness and honesty).

large—about .5. And people's guesses about the magnitude of that relationship are right on the money.

There is also pretty good recognition of the role of the law of large numbers in affecting correlations. If you look at scores summing across many behaviors and correlate them with the sum of another large batch of behaviors, the correlations are higher. People don't recognize how very much higher the correlation across summed behaviors is, but they do recognize that behavior over twenty occasions gives you a substantially better prediction for the next twenty occasions than behavior on one single occasion does for another single occasion.

Contrast the accuracy for abilities with the hopeless inaccuracy for traits. People think that honesty in one situation is correlated with honesty in another, and friendliness in one situation is correlated with friendliness in another, to the tune of .8! That is grievously wrong. The correlation between behavior on one occasion that reflects any personality trait whatsoever with behavior on another occasion reflecting that trait is typically .1 or less and virtually never exceeds .3. The error here is colossal and full of implications for everyday life that were discussed in the previous chapter. We think we can get a very good bead on someone's traits by observing their behavior in a single situation that taps that trait. This mistake is part and parcel of the fundamental attribution error, compounded by our failure to recognize that the law of large numbers applies to personality estimates just as it does to ability estimates. We think we learn much more than we do from a small sample of a person's behavior because we are inclined to underestimate the possible role of the context and because we think behavior on one occasion is sufficient to make a good prediction about behavior on the next, possibly quite different occasion. Moreover, there is virtually no recognition of the effect of increasing the number of observations. If you observe people's trait-related behaviors over a large number of occasions and correlate that total with the total of behaviors in another twenty situations, you do indeed get very high correlations. The problem is that people believe that the law of large numbers for observations of trait-related behavior also holds for a small number of observations of trait-related behavior!

Why is there such radically different accuracy at the level of single occasions measuring abilities and single occasions measuring traits? And why is there fairly decent recognition of the role of the law of large numbers in producing accurate measures for abilities but virtually no recognition at all for traits?

It's all in the *coding*. For many if not most abilities we know what the units are for measuring behavior and we can actually give them numbers: proportion of words spelled correctly; percentage of free throws made. But what are the appropriate units for judging friendliness? Smiles per minute? "Good vibes" per social exchange? How do we compare the ways that people manifest friendliness at Saturday night parties with the ways they show friendliness in Monday afternoon committee meetings? The types of behavior that people engage in are so different for the two types of circumstances that the things we're labeling as evidence of friendliness in one situation are quite different from what we're using as indicators of friendliness in another situation. And to try to give numbers to the friendliness indicators for situation A is difficult or impossible. Even if we could give numbers to them, we wouldn't know how to compare them to the numbers we have for the friendliness indicators for situation B.

What's the cure for the error with traits? We're not going to be able to define the relevant units of behavior with much accuracy and we're not going to give them numbers if we could. Psychologists do this in the context of research, but if we made such measurements, we couldn't mention it to a single soul because that person would think we were crazy. ("I'm giving Josh a score of 18 on friendliness of smiles at the meeting based on multiplying number of upward bends of the lips times the angle of each bend. Wait. Come back. Where are you going?")

The most effective way to avoid making unjustifiably strong inferences about someone's personality is to remind yourself that a person's behavior can only be expected to be consistent from one occasion to another if the context is the same. And even then, many observations are necessary for you to have much confidence in your prediction.

It may help to remember that *you* are not all that darned consistent. I'd bet that people who have met you in some situations have regarded you as pretty nice and people who have seen you in other situations have regarded you as not so nice at all. And I'd bet further that you couldn't blame people in those situations from reaching those conclusions given the evidence available to them. Just remember that it's the same for that guy you just met. You can't assume that you would experience his personality the same way in the next, perhaps rather different, situation in which you might encounter him.

More generally, know what you can code and what you can't. If you can't code or assign numbers to the event or behavior in question offhand, try the exercise of attempting to think of a way to code for it. The sheer effort it would take to do this is likely to alert you to the fact that you're susceptible to overestimating consistency of the event or behavior.

The best news I can offer you about the topics in this chapter and the preceding one is that, although I've shown how you can think statistically in just a tiny number of domains where you didn't previously, I know from my research on teaching people how to reason statistically that just a few examples in two or three domains are sufficient to improve people's reasoning for an indefinitely large number of events, even if they bear little resemblance to the ones I taught them about.

When I teach the law of large numbers with problems that people are inclined to reason about statistically anyway, such as lotteries and coin tosses, their inferences for the kinds of events they only sometimes think about probabilistically, such as objectively scorable abilities, improve.[15] Their inferences for the sorts of things they rarely think about statistically, such as personality traits, also improve. The same is true if I teach using just objectively scorable examples about abilities or teach using more subjective, difficult-to-score examples. Teaching about problems of one type improves reasoning about other, very different types.

Summing Up

Accurate assessment of relationships can be remarkably difficult. Even when the data are collected for us and summarized, we're likely to guess wrongly about the degree of covariation. Confirmation bias is a particularly likely failing: if some As are Bs, that may be enough for us to say that A is associated with B. But an assessment of whether A is associated with B requires comparing two ratios from a fourfold table.

When we try to assess correlations for which we have no anticipations, as when we try to estimate the correlation between meaningless or arbitrarily paired events, the correlation must be very high for us to be sure of detecting it. Our covariation detection abilities are very poor for events separated in time by more than just a few minutes.

We're susceptible to illusory correlations. When we try to assess the correlation between two events that are plausibly related to each other—for which we're prepared to find a positive correlation—we're likely to believe there is such a correlation even when there isn't. When the events aren't plausibly related, we're likely to fail to see a positive correlation even when a relatively strong one exists. Worse—we're capable of concluding there is a positive relationship when the real relationship is negative and capable of concluding there is a negative relationship when the real relationship is positive.

The representativeness heuristic underlies many of our prior assumptions about correlation. If A is similar to B in some respect, we're likely to see a relationship between them. The availability heuristic can also play a role. If the occasions when A is associated with B are more memorable than occasions when it isn't, we're particularly likely to overestimate the strength of the relationship.

Correlation doesn't establish causation, but if there's a plausible reason why A might cause B, we readily assume that correlation does indeed establish causation. A correlation between A and B could be due to A causing B, B causing A, or something else causing both. We too often fail to consider these possibilities. Part of the problem here is that we don't recognize how easy it is to "explain" correlations in causal terms.

Reliability refers to the degree to which a case gets the same score on two occasions or when measured by different means. Validity refers to the degree to which a measure predicts what it's supposed to predict. There can be perfect reliability for a given measuring instrument but no validity for the instrument. Two astrologers can agree perfectly on the degree to which Pisces people are more extroverted than Geminis—and there most assuredly is no validity to such claims.

The more codable events are, the more likely it is that our assessments of correlation will be correct. For readily codable events such as those determined by ability, our assessment of correlations across two occasions can be quite accurate. And we recognize that the average of many events is a better predictor of the average of many other events of the same kind than measurement of a single event is for another single event—when the events in question are influenced by some ability. Even for abilities, though, gain in predictability from observation on one occasion to predictability based on the average of many occasions tends to be substantially greater than we realize. Our assessments of the strength of relationships based on difficult-to-code events such as those related to personality can be wildly off the mark, and we show little or no recognition of the extent to which observations of many such events are a far better guide to future behavior than are observations of a few such events.

Caution and humility are called for when we try to predict future trait-related behavior from past trait-related behavior unless our sample of behavior is large and obtained in a variety of situations. Recognizing how difficult it is to code behavior of a particular kind may alert us to the possibility that our predictions about that kind of behavior are particularly susceptible to error. Reminding ourselves of the concept of the fundamental attribution error may help us to realize that we may be overgeneralizing.

PART IV

EXPERIMENTS

Inquiry is fatal to certainty.   —Will Durant, philosopher

Institutions increasingly rely on experiments to provide them with information. That's a good thing, because if you can do an experiment to answer a question, it's nearly always going to be better than correlational techniques. The correlational technique known as *multiple regression* is used frequently in medical and social science research. This technique essentially correlates many independent (or predictor) variables simultaneously with a given dependent variable (outcome or output). It asks, "Net of the effects of all the other variables, what is the effect of variable A on the dependent variable?" Despite its popularity, the technique is inherently weak and often yields misleading results. The problem is due to *self-selection*. If we don't assign cases to a particular treatment, the cases may differ in any number of ways that could be causing them to differ along some dimension related to the dependent variable. We can know that the answer given by a multiple regression analysis is wrong because randomized control experiments, frequently referred to as the gold standard of research techniques, may give answers that are quite different from those obtained by multiple regression analysis.

Even when assignment to condition is not literally random, we can sometimes find "natural experiments." These can occur when there happen to be groups of cases (people, agricultural plots, cities) that differ in interesting ways with respect to an independent variable, and there is no reason to assume that membership in the group is biased in some way that would prevent us from comparing the groups with respect to some dependent variable.

Society pays dearly for the experiments it could have conducted but didn't. Hundreds of thousands of people have died, millions of crimes have been committed, and billions of dollars have been wasted because people have bulled ahead on their assumptions and created interventions without testing them before they were put into place.

When it's human beings we're studying, there's a temptation to rely on verbal reports. Those reports are subject to a wide variety of errors. If we can possibly measure actual behavior rather than verbal reports, we're more likely to get a correct answer to a research question.

You can do experiments on yourself that will provide much more accurate answers about what affects your health and well-being than casual observation can produce.

9. Ignore the HiPPO

Shortly after Barack Obama announced he was running for president in the fall of 2007, Google's CEO, Eric Schmidt, interviewed him in front of a large audience of Google employees.[1] As a joke, Schmidt's first question was, "What is the most efficient way to sort a million 32-bit integers?" Before Schmidt could ask a real question, Obama interrupted: "Well, I think the bubble sort would be the wrong way to go," a response that was in fact correct. Schmidt slapped his forehead in astonishment, and the room broke out in applause. Later, in the question-and-answer period, Obama assured his audience, "I am a big believer in reason and facts and evidence and science and feedback," and promised that he would run the government accordingly.

In the audience that day was a product manager named Dan Siroker, who made a decision on the spot to go to work for Obama. "He had me at bubble sort."

Siroker had some science to offer Obama's campaign. He showed workers how to A/B test. When you don't know which of two treatments or procedures is best to achieve some goal, you compare the two by flipping a coin to decide who gets treatment A and who gets treatment B. You then collect data relevant to the question you're interested in and analyze the data by comparing the average of A with the average of B using a statistical test of some kind.

This chapter explains in detail what A/B testing is and how its principles can be applied in professional work and in your everyday life. If you understand how a good experiment is designed, you'll be better equipped to critique alleged scientific findings that you encounter in the media.

A/B

By the time Dan Siroker joined the Obama campaign website, developers for Google and other Internet companies had for several years been testing variations of web pages online. Instead of basing decisions about web design on HiPPOs—the derisive term for the "highest-paid person's opinion"—they were acting on incontrovertible facts about what worked best. A certain percentage of web users would be offered a home page design with lots of blue, and other users would be offered a design with lots of red. The information they sought was "percent who clicked." Potentially every aspect of the page from color to layout to images to text would be tested simultaneously on randomly selected users. The evidence, and not the HiPPO, was the decider about what should be on the website.

The application of A/B testing to political websites was straightforward. A major question was how to design a web page that would optimize the number of e-mail addresses for possible donors. For example, which button would get the most sign-ups—"Learn More," "Join Us Now," or "Sign Up Now"? Which image would get more sign-ups—a luminous turquoise photo of Obama, a black-and-white photo of the Obama family, or a video of Obama speaking at a rally?

I'm guessing you wouldn't have predicted that the combination of "Learn More" plus a family photo would be the most effective. And not just a little more effective. That combination produced 140 percent more donors than the least effective combination, which translates into a huge difference for donations and votes.

Website designers have learned what social psychologists discovered decades ago about their intuitions concerning human behavior in novel situations. As Siroker puts it, "Assumptions tend to be wrong."

From 2007 on, A/B testing dictated a wide range of Obama campaign decisions. The campaign specialist and former social psychologist Todd Rogers conducted dozens of experiments for Obama. Some of the experiments were shots in the dark. Is it better for donations and for voter turnout to get a robocall from Bill Clinton or a chatty call from a volunteer? (The latter, it turns out, by a lot.) A visit from a campaign worker just before Election Day is the single most effective way yet discovered to get someone to show up at the polls.

There is now a large body of research on what works for getting out the vote. Which is more effective at getting people to the polls: telling people that turnout is expected to be light or that turnout is expected to be heavy? You might think that telling people that voting is going to be light would make them more likely to vote. A quick cost-benefit analysis shows that your vote would count for more than if turnout was heavy. But remember how susceptible people are to social influence. They want to do what other people like them are doing. If most are drinking a lot, they'll go along; if they're not drinking a lot, they'll cut back. If most people are reusing their towels in their hotel room, so will they. And so telling voters there will be a heavy turnout in their precinct turns out to be much more effective than saying there will be a light turnout.

Is it effective to let people know that you know they voted in the last election—and that you'll be checking on them after this one? People want to look good in others' eyes—and in their own. So it's not surprising to learn that the promise of checking up can be worth 2.5 turnout percentage points or even more.[2] But only A/B testing could show whether the tactic of checking up would produce positive or negative results, or would have no impact at all.

In both 2008 and 2012 the Obama campaign had so many tricks up its sleeve that the Republican campaigns were blindsided. The Romney campaign in 2012 was so confident of victory that no concession speech was prepared for the candidate.

Republicans, however, are perfectly capable of playing the A/B game themselves. Indeed, already in 2006, the campaign of Governor Rick Perry of Texas had established that bang for the buck was poor for direct voter-contact mail, paid phone calls, and lawn signs. So the campaign spent no money on those things. Instead, the campaign used TV and radio spots heavily. Just which spots were the most effective was established by isolating eighteen TV markets and thirty radio stations and assigning start dates at random. Opinion polls tracked just which spots produced the biggest shifts toward Perry. The design's randomized nature added hugely to the accuracy of the results. Campaign workers weren't allowed to pick which market got which treatment at which time. If they had, any improved poll results could have been due to changed conditions in a given market rather than whether an ad had been placed in that market.

A/B testing can be just as useful for business as for politics, because researchers can segment the population and assign different treatments at random. When the number of cases (N) is very large, even very small differences can be detected. And in business as in politics, a small increment can make all the difference to success.

Doing Good While Doing Well

Merchandisers are making ever greater use of A/B testing. And they're finding that it can be as useful for finding ways to improve people's lives as it is for improving the bottom line.

Researchers in an El Paso, Texas, supermarket conducted A/B tests on a large number of tactics for increasing sales of fruits and vegetables.[3] Placing a divider in the shopping cart with a sign saying "Please place fruits and vegetables in the front of the cart" can double sales of fruits and vegetables, which are more profitable for the store than most other food items and will also have a beneficial effect on customer health.[4] Researchers have also put social influence to work. Signs telling customers that the average shopper in the store purchases X number of produce items can boost produce sales. And the signs turn out to have a massive effect on the purchases of the group with the most to gain by increasing their fruit and vegetable consumption, namely low-income people, many of whom are otherwise particularly likely to purchase processed foods and unlikely to buy fresh produce.

American grocery stores tend to group items by category: starches in aisle 4, sauces in aisle 6, cheeses in aisle 9. Japanese grocery stores tend to group items holistically, by type of meal: pasta, sauce, and cheese in the Italian meal section; tofu, seafood, and soy sauce in the Japanese section. The holistic technique can reduce processed food purchases and make customers who are pressed for time more likely to buy wholesome foods for home cooking.[5]

Organizations can conduct far more experiments on the effectiveness of their operations and their work environments than they do. Are employees more productive when they're allowed to work at home part of the time? All of the time? None of the time? Are high school students more likely to do their homework if one large assignment is given once a week or a small assignment is given every day?

Within Designs Versus Between Designs

A nationwide chain like Sears can randomly direct ads to particular segments of the public in a given media market, and it can randomly select where to put items of a given type in the store—in the back in New Hampshire and North Carolina and in the front in Vermont and South Carolina, for example. The number of Sears stores nationally is sufficiently large that A/B tests can have substantial *power*. The power of a statistical test is its ability to detect whether a difference of a given size is significant. The larger the N, the more confident you can be that a difference of a given size is real, and not due to chance.

You can increase power still further by using a *within design*, for example by flipping locations of products in the same store. This controls for the host of differences that can exist *between* stores. A typical within design is the *before-after* design. What are sales like when you put jewelry near the front of the store and undergarments near the back? What are they like when you reverse that? A/B tests with before/after designs are much more sensitive than simple A/B designs because you can get a "difference score" for each case and use that as your measure. This index compares sales at the Houston location before the treatment minus sales at the Houston location after the treatment. You're then looking at a score that controls for everything that can differ across locations and customer types: size and attractiveness of store, local customer preferences, and so on. Those sorts of differences are referred to as *error variance*, because they reflect variation between stores or between people that have nothing to do with the intervention: Scores may be high or low for reasons that are irrelevant to the question that A/B testing is intended to answer. You're more likely to know whether a difference in sales under condition A versus condition B is real or not when you reduce error variance by getting a before-score and an after-score for each case.

Note that if you're using a before/after design, you have to *counterbalance* the order of treatments. That is, some cases are exposed to the experimental condition first and some to the control condition first. Otherwise, treatment effects and order effects are *confounded*. What you think is an effect of the treatment may actually be an effect of order of events or simply of time.

Some before/after experiments occur by accident and yield serendipitous but useful results. My favorite such case concerns a southwestern gift shop.[6] Turquoise jewelry had been selling poorly, so one evening before the owner left for a brief trip, he decided to put the jewelry on sale and left a sign for his assistant reading "all turquoise in this case X 1/2." When the store owner returned, almost all the jewelry was gone. He was pleased to see that, but not nearly as pleased as when the assistant told him of his amazement that the jewelry had sold better at twice its regular price than it had before the markup! The assistant had misread the note to indicate that there should be a markup of 100 percent rather than a reduction by 50 percent.

Normally, price is a fairly good heuristic indicating value, so customers took the high price of the jewelry as an indication of its worth. That wouldn't work with all kinds of merchandise, of course, but turquoise is particularly susceptible to reliance on the cost cue because few people can make a knowledgeable assessment of its quality.

The power of the before/after design means that we can do genuine experiments on ourselves. Occasional acid indigestion but don't know exactly why? Keep a log of food and drink every day with special attention to likely culprits such as alcohol, coffee, soda, and chocolate. Then conduct an actual randomized experiment—flip a coin to decide whether to have a cocktail. And vary one thing at a time to avoid confounding variables. If you stop eating chocolate *and* drinking soda and your reflux improves, you won't know whether it's the food or the beverage that's the guilty party. Chapter 12 on verbal report, after considering some more scientific methodology, offers lots more suggestions about experimenting on yourself.

Statistical Dependence and Independence

A larger number of cases, and random assignment of cases to experimental condition, increase our confidence that a given effect is real. But another factor is just as crucial, namely what can count as a case. Suppose you try procedure A in classroom 1 with thirty students. Procedure A might be the standard way of teaching—lectures in class and homework to be done outside of class. And you try unconventional procedure B—lectures to be seen at home on video and guided "homework"—in classroom 2 with twenty-five students. What is the total number of cases (N)? Alas, not fifty-five, which would be a respectable number likely to show a significant difference if there really is a difference.

The N is 2. This is because N equals the number of cases only when there is *independence of observations*. But in the case of a classroom of students or any group of people who interact with one another during the period of the treatment and measurement of its effects, the individual behaviors are not independent of one another. Joan's confusion may rattle others; Billy's antics may lower everyone's score on the test. The behavior of every individual is potentially *dependent* on the behavior of every other individual. In such situations there is no ability to conduct significance tests unless the number of *groups* is fairly large, and in that case the N is the number of groups, not the number of individuals.

If you can't conduct statistical tests, there is inevitable uncertainty about exactly what was the effect of the different treatments. However, doing next time around whatever worked better the first time around is better than just relying on your assumptions.

The concept of independence is crucial to understanding a limitless range of events. Incredibly, as of 2008, financial ratings services such as Standard & Poor's used models of probable defaults for mortgages that assumed defaults were independent of one another.[7] Joe Doakes's default in Dubuque was assumed to mean nothing about the likelihood of Jane Doe defaulting in Denver. This is not completely unreasonable in normal times. But under a wide range of circumstances, including most certainly during a period when you're in the midst of rapidly rising home prices, you have to assume that there is a possibility that you are in a bubble. In that case, the likelihood of default for mortgage 20031A is statistically dependent on whether mortgage 90014C defaulted.

The rating agencies were not—and are not—disinterested parties. They're paid by the banks to do their ratings, and a rating company's services are more in demand if the company is in the habit of rating securities as safe. So whether the ratings agencies were unbelievably inept in creating their default models or were simply guilty of fraudulent practice, I'm not in a position to know. But in any case the lesson is clear: faulty scientific methodology can have catastrophic consequences.

Summing Up

Assumptions tend to be wrong. And even if they didn't, it's silly to rely on them whenever it's easy to test them. A/B testing is child-simple in principle: create a procedure you want to examine, generate a control condition, flip a coin to see who (or what) gets which treatment, and see what happens. A difference found using a randomized design establishes that something about the manipulation of the independent variable has a causal influence on the dependent variable. A difference found by using correlational methods can't guarantee that the independent variable actually exerts an effect on the dependent variable.

Correlational designs are weak because the researcher hasn't assigned the cases to their condition. For example, lots of homework versus little, radio ads versus circulars, high income versus low income. If you don't randomly assign cases—people, or animals, or agricultural plots—to a condition, you invite on board all kinds of uncertainties. Cases at one level of the independent variable may differ from those at another level in any number of ways, some of which can be identified and some of which can't. Any of the measured variables, or variables not measured or even conceived of, could be producing the effect rather than the independent variable of interest. And it might even be that the variable presumed to be dependent is actually producing differences in the variable presumed to be the independent one.

The greater the number of cases—people, agricultural plots, and so on—the greater the likelihood that you'll find a real effect and the lower the likelihood that you will "find" an effect that isn't there. If a difference is shown by a statistical test of some sort to be of such a magnitude that it would occur less than one time in twenty by chance, we say it's significant at the .05 level. Without such a test, we often can't know whether an effect should be considered real.

When you assign each case to all of the possible treatments, your design is more sensitive. That is to say, a difference of a given magnitude found by a "within design" is more likely to be statistically significant when tested in a "between" design. That's because all the possible differences between any two cases have been controlled away, leaving only the treatment difference as the possible cause of the relationship.

It's crucial to consider whether the cases you're examining (people in the case of research on humans) could influence one another. Whenever a given case might have influenced other cases, such that any one case could have had an impact on other cases, there's a lack of statistical independence. N is the number of cases that can't influence one another. Classroom A has an N not of the number of children in it but of just 1. (An exception would exist if influence could safely be considered to be minimal or nonexistent, such as when students take an exam in a room with cubicles where there is no talking.)

10. Experiments Natural and Experiments Proper

Since newborns have immature immune systems, every effort should be made to minimize their contact with bacteria and viruses that cause diseases.

—From *Germ Fighting Tips for a Healthy Baby*, CNN TV News,
February 2, 2011 (CNN, 2011)

Infants who come into contact with a wide range of bacteria very early in life appear to be at a lower risk of developing allergies later in life . . .

—From *Infants' Exposure to Germs Linked to Lower Allergy Risk*,
Canadian TV News, November 3, 2011 (CTV, 2011)

Friends, colleagues, and the media bombard us with advice about how to live our lives and conduct our professional activities.

Last decade we learned that fat in the diet should be minimized; now we're told that moderate quantities of fat are good for us. Last year's report maintaining that vitamin B6 supplements improve mood and cognitive function in the elderly is contradicted by this year's advice that it has no value for either outcome. Fifteen years ago a glass of red wine a day was supposed to be good for cardiovascular health; eight years ago alcohol in any form did the same thing; last week we're back to just red wine again.

Even if we're willing to default to accepting whatever is the latest medical advice, we have to reckon with conflicts between claims. Cousin Jennifer's dentist recommends twice-daily flossing and your own dentist advises that occasional flossing is adequate.

The New York Times's financial advice writer tells you to dump stocks and buy bonds. *The Wall Street Journal*'s guest columnist's advice is to buy real estate and park a lot of money in cash. Your financial advisor recommends a heavy allocation to commodities. Your friend Jake's financial advisor urges shifting funds from domestic to foreign stocks.

Your friends Eloise and Max are extremely anxious about getting their child into the best possible pre-K child care at whatever cost; your friends Earl and Mike think that the stimulation their toddler gets at home renders outside intellectual stimulation superfluous and are concerned only about a pleasant play environment.

This chapter gives tips about how to appraise scientific evidence presented in the media and proffered by acquaintances, and it makes suggestions about how to collect and assess evidence for yourself. You'll also read about how disastrous it can be when societies decide to rely on assumptions about the effects of interventions rather than conducting experiments about their effects.

A Continuum of Convincingness

So you watched CNN in February and they told you to keep your baby away from germs. Then you watched CTV in November and they told you that germs were good for your baby because they decrease the likelihood of autoimmune diseases such as allergies. Whom should you believe? What kind of evidence would make you choose to expose your baby to germs and what kind would make you inclined to keep your baby as far away from germs as possible? Here are some *natural experiments* you might find helpful in answering this question. A natural experiment allows a comparison between two (or among several) cases that are generally similar but differ in some way that might be related to an outcome variable of interest. No one manipulates that possibly relevant difference; if that were the case, it would be a genuine experiment. At the same time, at least we have no reason to assume that cases differ in some way that would render comparison meaningless.

Suppose you knew that East Germans were less likely to have allergies than West Germans.

Suppose you knew that Russians were less likely to have allergies than Finns.

Suppose you knew that farmers were less likely to have allergies than city dwellers.

Suppose you knew that children who had attended day care were less likely to have allergies than children who didn't.

Suppose you knew that children who had pets when they were infants were less likely to have allergies than children who didn't.

Suppose you knew that children who had lots of diarrhea as infants were less likely to have allergies than children who had less diarrhea.

Suppose you knew that infants who were born vaginally were less likely to have allergies than infants who were born by Caesarian section.

As it happens, all those things are true.[1] These natural experiments resemble true experiments in that similar cases happen to differ in some particular way (the independent variable, in effect) that might cause a difference in the outcome in question (the dependent variable of allergy). Each of the natural experiments provides a test of the hypothesis that early exposure to bacteria confers resistance to allergy, as well as to other autoimmune diseases such as asthma. (Autoimmune diseases are an abnormal, mistakenly "protective" overreaction against substances normally present in the body, in which white blood cells attack actual body tissues.)

Allergies range from annoying to debilitating in their effects, and asthma can be lots worse than that. Every day in the United States tens of thousands of kids miss school because of asthma, hundreds of people enter the hospital because of it, and some people die of it.

We can assume that East Germany and Russia are less hygienic places than West Germany and Finland, or at least were in the not too distant past. (Interestingly, many years ago a Polish immigrant to the United States told me half-jokingly that he thought allergies were an American invention. He may have been onto something.)

We can also assume that children raised on farms are more likely to be exposed to a wide variety of bacteria than children who grow up in cities. We know that children with pets are exposed to a wider range of bacteria—including fecal bacteria—than those who don't have pets. We know that toddlers are walking petri dishes who expose one another to a wider range of bacteria in pre-K programs than they would encounter if they just stayed at home. Lots of diarrhea can be the result of exposure to lots of bacteria. Vaginal birth exposes the infant to the full panoply of bacteria in the mother's vagina. These natural experiments all support the conclusion that bacteria are good for babies.

I doubt that these findings would encourage you to let your baby get down and dirty—to be exposed even to the yuckiest kind of bacteria associated with mucus and animal feces.

But what if you knew that swabs of babies' rectums that revealed a large variety of bacteria predicted lower autoimmune deficiency at the age of six? That happens to be the case.[2] We now have *correlational evidence*, or what is sometimes called *observational evidence*. Within a given population, the greater the exposure to a wide range of bacteria early on, the lower the incidence of autoimmune disease.

If you're still not ready to expose your baby to large amounts of diverse germs, it might influence you to know that there is a fairly plausible hypothesis called the "germ exposure theory" that could account for the correlational and natural experiment evidence. Early exposure to bacteria could be expected to stimulate the immune system and such stimulation might have beneficial effects down the road. The young immune system might be strengthened in ways that allow it to adapt and regulate itself, with the result that there is less inflammation and less susceptibility to autoimmune disorders later.

Now are you ready to get your baby dirty? Personally, I'm not sure I would be. Natural experiments, correlational evidence, and plausible theories are all well and good. But I would want to see a *true experiment* of the *double-blind, randomized control* sort, with babies assigned by the proverbial flip of a coin to an *experimental* high-bacteria exposure condition versus a *control*, low-bacteria condition. Both the experimenter and the participants (the mothers in this case) should be ignorant of (blind to) the condition the babies were assigned to. Ignorance resulting from this double-blind design rules out the possibility that results could have been influenced by either the experimenter's or the participant's knowing what condition the participant was in. If it turned out that the experimental high-exposure babies had fewer problems with allergy and asthma, I would be willing to seriously entertain allowing my baby to be exposed to a wide range of bacteria.

But I'm not quite sure I would be willing to allow my baby to be a guinea pig in the experiment that might convince me. Fortunately no one has to volunteer her baby for that experiment. There are such things as experiments with *animal models*. This is a living animal, phylogenetically close to humans, for which treatments can be presumed to have effects similar to those we would find for humans.

Researchers have studied the effects of bacteria exposure in young mice.[3] Instead of exposing some of the mice to extra-high levels of bacteria, the scientists went the other direction and created a germ-free environment for some of the mice and left control mice in normal lab-mouse conditions, which, trust me, are decidedly not germ-free. The germ-free mice developed abnormal levels of a type of killer T cell in parts of the colon and in the lungs. These T cells were subsequently recruited to attack even nonthreatening substances, with resulting inflammation, allergy, and asthma.

I guess I would now side with the CTV recommendation to let my kid wallow in dirt. Though admittedly it would make me extremely nervous. (And be wary of my advice here. Remember I'm not a real doctor, as my son occasionally reminds me, but only a PhD.)

If you decide to expose your infant to extra doses of bacteria, note that such exposure seems to have benefits primarily in the first couple of years of life. So you may not want to continue deliberate exposure to microbes indefinitely.

Believe it or not, the week I finished writing the above paragraphs, an article appeared in *JAMA Pediatrics* showing that infant colic, believed by some to be due to irritable bowel syndrome, is greatly alleviated by the administration of five drops of a solution containing the bacterium *Lactobacillus reuteri*.[4] The treatment produces a nearly 50 percent reduction in infants' colic-related crying.

What if your young child does get an infection? Should you follow the doctor's recommendation to give the child an antibiotic?

What if you knew that the richer the country, the higher the rate of inflammatory bowel disease (IBD), including Crohn's disease and ulcerative colitis?[5] These diseases can be very serious, even fatal. They can cause abdominal pain, vomiting, diarrhea, rectal bleeding, severe internal cramps, anemia, and weight loss. It should raise your suspicions to know that IBDs, like allergy and asthma, are autoimmune diseases. Circumstantial evidence of a correlational nature, to be sure. Surely wealth per se couldn't cause inflammatory bowel disease.

But something associated with wealth could be creating problems. People of a certain age will remember that their childhood was plagued with middle-ear infections whereas their own children, courtesy of amoxicillin, got rid of them almost as soon as they got them. The richer the country, of course, the more likely you are to visit a doctor, to have antibiotics prescribed, and to be able to fill the prescription with insurance money or your own money.

But if you're like me, you may have wondered whether it was a great idea to get all those antibiotics. And it looks like I was right to worry. Kids who have lots of ear infections and lots of antibiotics are more likely to develop IBD later in life.[6]

Antibiotics are overeager. They kill the good, the bad, and the ugly among the microorganisms in the gut.

Antibiotic use even in adulthood appears to be associated with subsequent bowel disease. Researchers have found that adults who have IBD are twice as likely to have had multiple antibiotic prescriptions in the previous two years.[7]

Our evidence is still circumstantial. What's needed is a true experiment. As it happens, the right one exists.

If lack of good bacteria is the problem causing IBD, then infusions of good bacteria into the gut, for instance, via an enema containing some of the contents of a healthy person's intestines, should be an effective treatment for IBD.

Brave scientists, and even braver patients, tried the experiment. ("So, Ms. Jones, in this experiment we're going to pump the contents of a stranger's intestines into yours. And not just because it's never been done, but because there's a possibility it may be good for you.") Lucky for both the patients and the scientists, the experiment worked, and treated patients were more likely to improve than were control subjects who received only saline solution. (And lucky for you, it's now possible to purchase the helpful intestinal bacteria in pill form.)

A decision about treating any given childhood disease with antibiotics awaits a lot of research and a thorough cost-benefit analysis, of course. The same goes for infections people develop in adulthood.

From Natural to Proper Experiments

Natural experiments can have tremendously important implications that beg to be studied by proper experiments.

Children of parents with little education, and who are therefore at risk for low academic achievement themselves, are likely to have a poor elementary school outcome if their first-grade teacher, judged by observers, is in the bottom third of teaching effectiveness. If they're lucky enough to get a teacher in the top third of effectiveness, their performance is likely to nearly equal the performance of middle-class children.[8] This finding constitutes a natural experiment. If children were to be randomly assigned to classrooms with teachers of different judged competence, we would have a true experiment. Meanwhile, what parent would be indifferent to teacher effectiveness after hearing about the result of the natural experiment?

Greenery in a city is nice. Nicer than you might assume, actually. A study of identical public housing apartment dwellings in Chicago found about half as many reported crimes in apartment houses surrounded by greenery as in apartment houses surrounded by barren land or concrete.[9] In light of the sort of subtle situational cues that can profoundly affect behavior, discussed in Chapter 1, this is not such a surprising finding. The study is probably a real experiment because housing officials in Chicago believe that assignments to a particular project are made randomly—and there's no reason to assume that's not correct. On the other hand, laypeople don't necessarily mean the same thing by the term "random" as scientists do, so complete confidence in the greenery/low crime hypothesis awaits a study with verifiably random assignment to dwelling to rule out the possibility that the relationship between greenery and crime is causal and not merely correlational. Obviously, such an experiment is sorely needed. If the true experiment findings duplicate the natural experiment findings, a cost-benefit analysis of the kind discussed in Chapter 4 would be badly needed. Such a study would examine the effects of ripping up concrete and putting in trees and weigh them against the cost in dollars. The analysis might show that the landscape change is a bargain for a city.

Scientists often get their ideas when they realize that some observation they've made constitutes a natural experiment. The eighteenth-century physician Edward Jenner noticed that milkmaids rarely got smallpox, a disease related to the cowpox to which the milkmaids would have been exposed. Maybe milkmaids were less likely to get smallpox than butter churners because cowpox somehow protected against smallpox. Jenner found a young milkmaid with cowpox on her hand and inoculated an eight-year-old boy with some material from it. The boy developed a fever and discomfort in his armpits. A few days later, Jenner inoculated the boy with smallpox from a lesion from a smallpox sufferer. The boy did not develop the disease, and Jenner correctly announced that he had discovered a treatment that could prevent smallpox. The Latin word for cow is *vacca*, and Latin for cowpox is *vaccinia*, so Jenner called his treatment *vaccination*. A natural experiment led to a proper experiment, and the results changed the world for the better. Smallpox exists today only in the form of a single virus specimen kept alive in a single laboratory. (It's being kept because there would be a need for a source for vaccination material if the disease were to crop up somewhere in the world.)

The High Cost of *Not* Doing Experiments

We can pay dearly, in blood, treasure, and well-being, for experiments that aren't done.

In the nearly fifty years that Head Start has been in existence, we have spent $200 billion on it. Head Start is a preschool program for poor, primarily minority children intended to improve their health, academic achievement, and, it was hoped, their IQs. What have we gotten for our investment? The program did improve the children's health, and initially improved IQ and academic success. But the cognitive gains lasted only a few years; by mid–elementary school the children were doing no better than children who hadn't been in the program.

We don't know for sure whether the Head Start children fared any better as adults than did children who weren't in the program.[10] That's because assignment to the program was not random. Kids who ended up in Head Start could have differed in any number of unknown ways from those who didn't attend the program. All the adult outcome data, of which there is shockingly little, rely on purely *retrospective* information about assignment. People had to remember whether they had been in a preschool program and if so, which one. Retrospective studies are subject to a great deal of potential error, especially when the memories in question go back to events decades in the past. The retrospective studies do show apparent gains in life outcomes in adulthood for children who were in Head Start.[11] But this result doesn't even come up to the level of a natural experiment because it would be surprising if there weren't preexisting differences between children who were in Head Start and those who were not.

A lot of money continues to be spent on something that may or may not be effective.

Fortunately, as you'll recall from Chapter 4, we know some preschool programs have a huge effect on adult outcomes. Randomized-assignment experiments with programs more intensive than Head Start produced modest IQ gains that were long-lasting, but, much more important, academic improvements and economic gains for adults who had been in the treatment groups were huge.

The costs of not knowing what works and what doesn't in the way of preschool programs have been very great indeed. The $200 billion for Head Start might have been better spent on a smaller number of particularly vulnerable children, providing them with more intensive experiences. That might have produced far greater societal benefits. (And we do in fact know that the poorer the child, the greater the impact of high-quality early childhood education. It doesn't seem to much affect outcomes for middle-class children.)[12] Moreover, no experiments were conducted to find out what aspects of Head Start (if any) were the most effective. Is it better to focus on academics or on social factors? Half days or full days? Are two years needed or would one year make almost as much difference? The social and economic consequences of knowing the answers to such questions would be enormous. And getting the answers would have been easy and dirt cheap in comparison to what has been spent.

At least it's unlikely that Head Start does any harm to children who participate in it. But many interventions dreamed up by nonscientists actually do harm.

Well-intentioned people invented a program to help possible trauma victims soon after a tragedy has occurred. So-called grief counselors encourage participants in a treatment group to recount the incident from their own perspective, describe their emotional responses, offer their comments on others' reactions, and discuss their stress symptoms. The counselor assures participants that their reactions are normal and that such symptoms generally diminish with time. Some nine thousand grief counselors descended on New York City in the wake of 9/11.

Grief counseling of this sort seems like an excellent idea to me. However, behavioral scientists have conducted more than a dozen randomized experiments examining critical incident stress debriefing (CISD). They have found no evidence that the activity has a positive effect on depression, anxiety, sleep disturbance, or any other stress symptoms.[13] There is some evidence that people who undergo CISD are *more* likely to experience full-blown traumatic stress disorder.[14]

As it happens, behavioral scientists have found some interventions that actually are effective for trauma victims. A few weeks after a critical incident, the social psychologist James Pennebaker has trauma victims write down, in private, and for four nights in a row, their innermost thoughts and feelings about the experience and how it affects their lives.[15] And that's all. No meetings with a counselor, no group-therapy encounters, no advice about how to handle the trauma. Just a writing exercise. The experience typically has a very substantial effect on suffering from grief and stress. It's not at all plausible to me that this exercise would be very effective. Certainly not as plausible as the idea that immediate intervention, grief sharing, and advice would be effective. But there it is. Assumptions tend to be wrong.

Pennebaker thinks his writing exercise works because it helps people, after a period of suffering and incubation, to develop a narrative to understand the event and their reactions to it. And it seems to be the case that the people who improve most are those who began the exercise with inchoate and disorganized descriptions and ended with coherent, organized narratives that gave meaning to the event.

Other well-meaning people have tried to inoculate teenagers against peer pressure to commit crimes and engage in self-destructive behavior, with results that are sometimes even more disappointing than CISD for trauma victims.

Decades ago inmates in Rahway State Prison in New Jersey decided to do something to warn at-risk adolescents of the dire consequences of criminal behavior. The inmates showed the kids what prison was like, including graphic accounts of rape and murder within its walls. An award-winning documentary on the Arts and Entertainment (A&E) channel christened the program Scared Straight. The name and the practice spread widely throughout the United States.

Do Scared Straight programs work? Seven experimental tests of the programs have been carried out.[16] Every single study found the Scared Straight kids to be *more* likely to commit crimes than kids in the control group who were exposed to no intervention at all. On average, the increase in criminal activity was about 13 percent.

The Rahway program still exists, and to this point more than fifty thousand East New Jersey kids have passed through it. Let's multiply fifty thousand by 13 percent. The figure we get is sixty-five hundred. That's how many more crimes have been committed than would have been if the well-meaning convicts had never thought up their scheme. And that's just one area of New Jersey. The program has been duplicated in many other communities. A study commissioned by the Washington State Institute for Public Policy estimated that every dollar spent on Scared Straight incurs crime and incarceration costs of more than two hundred dollars.

Why doesn't Scared Straight work? It certainly seems to me that it should. We don't know why it doesn't work, and we certainly don't know why it should be counterproductive, but that doesn't matter. It's a tragedy that it was invented and a crime that it hasn't been stopped.

Why hasn't it been stopped? I'll venture the guess that it just seems so obvious that it should work. Many people, including many politicians, prefer to trust their intuitively compelling causal hypotheses over scientific data. It doesn't help that scientists can't offer any convincing explanations for why Scared Straight doesn't work. Scientists, especially social scientists, don't fall into the trap of holding on to their intuitive causal theories in the face of conflicting data, because they are well aware that ATTBW: Assumptions Tend to Be Wrong. (As of this writing, the A&E channel is still airing a program singing the praises of Scared Straight.)

D.A.R.E. is another elaborate attempt to keep kids out of trouble. As part of the Drug Abuse Resistance Education program, local police officers undergo eighty hours of training in teaching techniques and then visit classrooms to present information intended to reduce drug, alcohol, and tobacco use. It's been funded by state, local, and federal government sources to the tune of $1 billion per year. According to D.A.R.E.'s website, 75 percent of American school districts participate in the program as well as forty-three countries.

But in fact D.A.R.E., as it has been conducted for the past thirty years at least, doesn't decrease children's use of drugs.[17] D.A.R.E. doesn't admit the ineffectiveness of its programs and actively combats critics who present scientific evidence for its failures. Programs intended by D.A.R.E. to supplement or replace the original have not been thoroughly evaluated by external institutions to this point.

Why doesn't D.A.R.E. work? We don't know. It would be nice if we did, but causal explanations are unnecessary. As it happens, some programs intended to lower the likelihood of drug, alcohol, and tobacco use *do* work. These include LifeSkills Training and the Midwestern Prevention Project.[18] These programs have elements missing from the original D.A.R.E. program, notably teaching preadolescents skills in resisting peer pressure. The inventors of D.A.R.E. made the assumption that police are important social influence agents for teenagers. A social psychologist could have told them that peers are a much more effective source of influence. The more successful programs also provide information about drug and alcohol use among teenagers and adults. Recall that such information often surprises because these rates are lower than most youngsters believe, and accurate knowledge about others' behavior can lower rates of abuse.[19]

Meanwhile, programs that damage young people are still being conducted, and programs that help are underused or used not at all. Society is paying a high price in dollars and human suffering for wrong assumptions.

Summing Up

Sometimes we can observe relationships that come close to being as convincing as a genuine experiment. People whose childhoods were spent in circumstances that would have resulted in relatively great exposure to bacteria are less prone to some autoimmune diseases. When this is found across a large number of quite different circumstances—hygienic versus less hygienic countries, farms versus cities, pets versus no pets, vaginal versus Caesarian birth, and so on—the observations begin to be very suggestive. Such observations led scientists to conduct actual experiments that established that early exposure to bacteria does in fact reduce the likelihood of autoimmune diseases.

The randomized control experiment is frequently called the gold standard in scientific and medical research—with good reason. Results from such studies trump results from any and all other kinds of studies. Randomized assignment ensures that there are no differences in any variable between experimental and control cases prior to the manipulation of the independent variable. Any difference found between them can usually be assumed to be due only to the scientist's intervention. Double-blind randomized control experiments are those where neither the researcher nor the patient knows what condition the patient is in. This type of experiment establishes that only the intervention, and not something about the patients' or doctors' knowledge of the intervention, could have produced the results.

Society pays a high cost for experiments not carried out. Because of failure to carry out randomized experiments, we don't know whether the $200 billion paid for Head Start was effective in improving cognitive abilities or not. Because of randomized control experiments, we do know that some high-quality pre-K programs are enormously effective, resulting in adults who function in much healthier and more effective ways. Proper experiments on pre-K techniques stand a chance of resulting in huge cost savings and great benefits to individuals and society. D.A.R.E. programs don't produce less teen drug or alcohol use, Scared Straight programs result in more crime, not less, and grief counselors may be in the business of increasing grief rather than reducing it. Unfortunately, in many domains, society has no means of ensuring that interventions are always tested by experiment and no way of guaranteeing that public policy must take into account the results of experiments that are carried out.

11. Eekonomics

Do auto salespeople offer deals to women that are more expensive than the deals they offer to men?

Does classroom size affect learning?

Are multivitamins good for your health?

Is there employer prejudice against the long-term unemployed—simply because they've been out of a job for a long time?

Should postmenopausal women take hormone replacement therapy to reduce the likelihood of cardiovascular disease?

Many answers to each of these questions have been proposed. Some answers were based on studies that reached incorrect conclusions because of faulty methodology. Some answers are quite likely correct because good scientific methodology was used.

This chapter makes three points that are crucial to understanding scientific findings and deciding whether to believe them.

1. Studies that rely on correlations to establish a scientific fact can be hopelessly misleading—even when the correlations come in complicated packages called "multiple regression analysis" that "control for" a host of variables.
2. Experiments in which people (or objects of any kind) are assigned randomly to one treatment versus another (or no treatment at all) are in general far superior to research based on multiple regression analysis.
3. Assumptions are so often wrong when it comes to human behavior that it's essential to conduct experiments if at all possible to test any hypothesis about behavior that matters.

Multiple Regression Analysis

All the questions that began this chapter ask whether some *independent* or *predictor variable*—an input or a presumed cause—affects some *dependent* or *outcome variable*—an output or an effect. Experiments manipulate independent variables; correlational analyses merely measure independent variables.

One technique employing correlational analysis is multiple regression analysis (MRA), in which a number of independent variables are correlated simultaneously (or sometimes sequentially, but we won't talk about that variant of MRA) with some dependent variable.* The predictor variable of interest is examined along with other independent variables that are referred to as *control variables*. The goal is to show that variable A influences variable B "net of" the effects of all the other variables. That is to say, the relationship holds *even* when the effects of the control variables on the dependent variable are taken into account.

Consider this example. Smoking cigarettes is correlated with higher incidence of cardiovascular disease. The temptation is to say it looks as if smoking *causes* cardiovascular disease. The problem is that lots of other things are correlated both with smoking and with cardiovascular disease, such as age, social class, and excess weight. Older smokers have smoked for a longer time than younger smokers, so we need to pull age out of the smoking-disease correlation. Otherwise, what we're showing is that being older *and* smoking are associated with cardiovascular disease. But that conflates two variables. What we want to know is just the association between smoking and cardiovascular disease, regardless of how old a person is. We "control for" age effects on cardiovascular disease by removing the age-disease correlation from the smoking-disease

---

*The term "regression" is a little confusing because "regression to the mean" seems to be a very different thing than examining the relation between a set of independent variables and a dependent variable. The reason the same word is used for such different purposes appears to be that Karl Pearson, the inventor of the correlation technique that bears his name, first used that method to examine the correlation between related individuals for some variable. The correlation between the heights of fathers and sons always shows a regression to the mean. Unusually tall fathers have somewhat shorter sons on average; unusually short fathers have somewhat taller sons on average. A correlation is a simple regression analysis relating one variable to another. Multiple regression examines the relation of each of a set of variables to another variable.

correlation. The result is that we can now say, in effect, that the association between smoking and cardiovascular disease is found for every age group.

The same logic applies to social class. Other things being equal, the lower the social class, the greater the likelihood of smoking, and the lower the social class, the greater the risk of cardiovascular disease, independent of any risk factor such as smoking. Ditto for excess weight. And so on. The correlations of these variables with both smoking and degree of cardiovascular disease have to be pulled out of the correlation between smoking and cardiovascular disease.

The theory behind multiple regression analysis is that if you control for everything that is related to the independent variable and the dependent variable by pulling their correlations out of the mix, you can get at the true causal relation between the predictor variable and the outcome variable. That's the theory. In practice, many things prevent this ideal case from being the norm.

First of all, how do we know we've identified all of the possible confounds—the variables linked to both the predictor and outcome variables? We're almost never able to make that claim. We can only measure what we assume might be important and leave out the infinity of variables that we don't assume are important. But ATTBW: Assumptions Tend to Be Wrong. So the battle is usually lost right there.

Second, how well do we measure each of the possible confounding variables? If we measure a variable poorly, we haven't controlled for it enough. If we measure a variable so poorly that it has no validity, we haven't controlled for anything.

Sometimes MRA is the only research tool available for examining interesting and important questions. An example is the question of whether religious belief and practice are associated with greater or lesser rates of procreation. We can't do an experiment to test that question, randomly assigning people to be religious or not. We can only use correlational methods such as MRA. As it happens, religiosity is correlated with fecundity at both the individual level and the national or cultural level. Controlling for income, age, health status, and other factors at the level of individuals, at the level of ethnic groups, and at the level of countries, greater religiosity is correlated with greater fecundity. Just why this is we don't know, and the correlation between religiosity and fecundity might not be causal at all but rather could be due to the fact that some unmeasured third variable influences both religiosity and fecundity. Causality could even run in the reverse direction: having lots of children might make people search for divine support and guidance! The correlational finding is interesting nonetheless, and knowing about it might have real-world consequences.

I want to be very clear that not all correlational research, or all multiple regression research, is valueless. I've often used multiple regression myself, even when I've conducted experiments establishing a causal relationship. I feel more comfortable knowing that a given relationship exists in the wild and not just in the laboratory or in a possibly atypical ecological environment.

Moreover, there are often clever things that one can do to make us pretty sure that we've learned something about causality. Take the correlation between the wealth of nations and the IQ of nations. What's going on causally there? The correlation taken by itself is hugely problematic. Lots of things are correlated with both wealth and IQ—physical health, for example. "Healthy, wealthy, and wise" is not just an expression; the three go together in a bundle of correlations that include many other potential causal variables as well. Moreover, there's a plausible causal story running in both directions. As a nation gets smarter it gets richer, because more advanced and complex ways of making a living become possible. As a nation gets richer it gets smarter, because wealth generally increases the quality of education.

But we can sometimes tell a pretty good causal story by looking at what are called "lagged correlations," that is, the correlation of an independent variable (assumed cause) with another variable (assumed outcome of the cause) at a later time. If a nation gets smarter—because of an increase in education, for example—does it get richer down the road? Indeed it does. For example, a few decades ago, Ireland made a concerted and highly successful effort to improve its educational system, especially at the high school, vocational school, and college levels.[1] College attendance actually increased by 50 percent over a brief period of time.[2] Within about thirty years, the per capita GDP of Ireland, which previously had IQ scores far lower than that of England (for genetic reasons, according to some English psychologists!), had exceeded the per capita GDP of England. Finland also made significant educational improvements beginning several decades ago, focusing especially on making sure the poorest students got an education as equal as possible to that of the richest students. By 2010, Finland was ahead of every country on international tests of academic achievement, and its per capita income had risen to be greater than that of Japan and Britain and only slightly less than that of the United States. Nations that have not made heroic efforts to improve education in recent decades, such as the United States, have declined in per capita income relative to other advanced countries. Such data are still correlational, but they indicate that as a nation begins to break out of the pack educationally, it begins to get richer. As it stagnates educationally, it begins to lose wealth relative to other nations. Pretty persuasive.

Many other circumstances can pull correlational research up to a level of convincingness equal to natural experiments or even randomized control experiments. For example, sheer magnitude of an effect can sometimes make us feel that it must not be a mere artifact due to correlated variables. We also become somewhat more confident that a given treatment is real if the effect is "dose-dependent." That is, the more intense or frequent the treatment, the higher the level of response. For example, people who smoke two packs a day are much more likely to have poor cardiovascular function than people who smoke half a dozen cigarettes a day. This makes it more likely that smoking really does worsen cardiovascular health than if it were the case that the sheer amount of smoking is unrelated to morbidity.

But there are serious problems with multiple regression analysis as it's all too frequently conducted. I'm going to be very explicit about the problems, because the media constantly report findings based on this highly fallible method, and important policy decisions are made on the basis of them. Epidemiologists, medical researchers, sociologists, psychologists, and economists all use this technique. It can produce serious errors, and the claims of some devotees that it can reveal causality are usually bogus.

In many instances, MRA gives one impression about causality, and actual randomized control experiments give another. In such cases we should believe in the results of the experiments.

Would you think the number of children in a classroom matters for how well schoolchildren learn? It seems reasonable that it would. But dozens of MRA studies by highly regarded investigators tell us that, net of average income of families in the school district, size of the school, IQ test performance, city size, and geographic location, average class size is uncorrelated with student performance.[3] The implication: we now know we needn't waste money on decreasing the size of classes.

But scientists in Tennessee conducted a randomized experiment in which they varied class size substantially. By the flip of a coin, researchers assigned children in kindergarten through third grade to either small classes (thirteen to seventeen) or larger classes (twenty-two to twenty-five). The study found that smaller classes produced about a .22 SD improvement in standardized test performance; the effect on minority children was greater than the effect on white children.[4] There are now three other experimental studies of the effects of reduction of class size, and their findings are almost identical to those of the Tennessee study.[5] These four experiments are not merely additional studies on the effects of class size. They *replace* all the multiple regression studies of class size. That's because we have much greater confidence in experimental results for a question like this.

Why do the multiple regression studies find that class size matters so little? I don't know, but we don't have to know in order to have a strong opinion about whether class size can matter.

There's a lot that the four experiments leave unanswered, of course. We don't know whether class size makes a difference for every region of the country, every degree of urbanization, every level of social class, and so on. We don't know what's going on in classrooms that produces the different educational effects. But those questions can be answered by further experiments, and positive findings for each experiment looking at populations that differ in notable ways from those examined in the available studies would increase our confidence that larger class size does indeed make a difference.

Whether decreasing the size of classes is the best thing to spend the educational dollar on is a separate question, and an answer to it is above my pay grade. Finland doesn't have particularly small classes; the improved educational outcome was more likely a result of paying teachers more and recruiting them primarily from the top of their college classes rather than the bottom, as the United States now does. And in any case, policy can't be determined simply by finding a beneficial effect of X on Y; a full-dress cost-benefit analysis is required.

The problem with correlational studies such as those based on MRA is that they are by definition susceptible to errors based on self-selection. Cases—people, classrooms, agricultural plots—differ in any number of ways. Longtime smokers aren't just longtime smokers, they've dragged along lots of other factors associated with smoking—such as greater age, lower social class, and excess weight. Classroom A is larger than classroom B, but it also differs in any number of ways over which the investigator has no control. Classroom A may have a better teacher because the principal thought that teacher could handle large-size classes better than other teachers in the school. Classroom B may have high achievement scores, even though it has a large number of students, because the principal thought that more able students would suffer less from relative lack of attention than less able students. And so on. The problem doesn't get solved simply by adding more classrooms, or more control variables, into the mix.

In studies where cases are randomly assigned to an experimental condition, the variability among classrooms on other dimensions remains. But—critically—it's the experimenter who selects the condition. This means that the experimental classrooms and the control classrooms have equally good teachers on average, equally able and motivated students, and equal resources. The classrooms have not "selected" their own level on each of these variables; the experimenter has. Thus the only thing that differs between experimental and control classrooms on average is the variable of interest, namely class size. Experiments like those on class size are not conclusive. Neither teachers nor administrators, for example, are blind to condition. They know which classes are small and which are large, and this might affect how teachers teach, including how much effort they put into the job. It's just that such problems pale in significance to the self-selection problem.

Medical Muddles

Did you know that consuming large amounts of olive oil can reduce your risk of stroke by 41 percent?[6] Did you know that if you have cataracts and get them operated on, your mortality risk is lowered by 40 percent over the next fifteen years compared to people with cataracts who don't get them operated on?[7] Did you know that deafness causes dementia? Did you know that being suspicious of other people causes dementia?

If those claims sound dubious to you, they should. But alleged findings such as these appear constantly in the media. They're typically based on epidemiological studies. (Epidemiology is the study of disease patterns in populations and their causes.) A great deal of epidemiological research relies on MRA. Studies using it attempt to "control" for factors such as social class, age, and prior state of health. But they can't get around the self-selection problem. The sorts of people who get a given medical treatment, or who consume large amounts of a particular food, or who take a particular vitamin, differ from those who don't get the treatment, don't consume the food, or don't take the vitamin, in goodness knows how many ways.

Let's look at the study claiming that, net of control factors including "socio-demographic variables, physical activity, body mass index, and risk factors for stroke," people who consume more olive oil have fewer strokes.[8] The reduction in one study was 41 percent for "intensive" users of olive oil versus those who never used it. But it might not be olive-oil consumption that reduces mortality but something correlated with olive-oil consumption. For starters, take ethnicity. Italian Americans are big olive-oil users, and African Americans almost surely are not. And the life expectancy of Italian Americans is significantly greater than for blacks, who, incidentally, are particularly prone to have strokes.

The biggest potential confound in any epidemiological study is typically social class. Class is a glaringly obvious candidate for affecting differences in risk for stroke and many if not most other medical outcomes. The rich are different from us. They have more money. People with more money can afford to use olive oil rather than corn oil. People with more money are also more likely to be widely read, and associate with other readers, and therefore to believe that olive oil is better for one's health than its cheaper competitors. People with more money get better medical care. And people with more money—and with higher social class, whether you measure class by education, personal income, or occupational status—have better life outcomes of all kinds.

A failure to control for social class in an epidemiological study is fatal to any attempt to infer the cause of a given medical outcome. But supposing the investigator does try to measure social class, how should it be done? Some use income, some use education, some use occupation. Which is best? Or should you somehow combine the three? The truth is that different epidemiological studies may try to measure social class in any or all or none of these ways. And that contributes to the constant churning of "medical findings" reported by the media.[9] (Fat is bad for you. No. Fat is good for you. Red meat is good for you. No. Red meat is bad for you. Antihistamines reduce the severity of the common cold. No. Antihistamines have no effect.) The differing results are frequently just the consequence of defining social class differently, or of not examining it at all.

But social class is only one of an unlimited number of potential confounds present in MRA studies. Almost anything that's correlated with both the predictor variable and the outcome variable in such studies becomes a candidate for explaining the correlation between the two.

There are thousands of food supplements on the market. MRA studies sometimes find that one or another supplement is beneficial for one thing or another. The media then pass along that finding. Unfortunately, there is usually no way for the reader to tell whether a given study is based on MRA, in which case you probably shouldn't pay much attention to it, or an actual experiment, in which case it may be very important indeed for you to pay attention. Reporters, even those who specialize in health reporting, typically don't fully understand the crucial difference between the two methodologies.

There are countless examples of MRA studies finding one thing and experiments finding another. For example, MRA studies report that vitamin E supplements reduce the likelihood of prostate cancer. By contrast, an experimental study was conducted at many locations in the United States, assigning at random some men to vitamin E supplements and others to a placebo. This experiment found a slight *increase* in the likelihood of cancer by virtue of taking vitamin E.[10]

Vitamin E is not the only suspect supplement. There are a host of experimental studies showing that taking multivitamins—something that almost half of Americans do—does little or no good, and very high doses of some vitamins can do genuine harm.[11] There is almost no evidence one way or another about the effects of any of the other fifty thousand or so food supplements that are on the market. Most of the evidence we do have about any given supplement indicates that it's useless; some indicates it's actually harmful.[12] Unfortunately, lobbying by the supplement industry resulted in Congress exempting supplements from federal regulation, including any requirement that manufacturers do experimental studies of actual effectiveness. As a consequence, billions of dollars are spent every year on nostrums that are useless or worse.

Using Multiple Regression Analysis When Only Experiments Can Do the Job

The longer a person is unemployed, the harder it is to find a job. As of this writing, the number of people who have been unemployed just for a short time (fourteen weeks or less) is only slightly more than it was just before the Great Recession hit.[13] But the number of long-term unemployed is 200 percent higher than it was then. Is there prejudice against the longtime unemployed on the part of employers? Are they not given consideration simply because they've been out of a job for a long time? MRA can't tell us whether, other things being equal, employers unjustifiably pass over the longtime unemployed in favor of the briefly unemployed. The long-term unemployed may have poor employment records, or be lackadaisical in job hunting, or be too picky about the kind of job they would do. Politicians routinely invoked these alleged causes during the Great Recession. But you can't know whether these explanations are correct by conducting a multiple regression analysis. No amount of "controlling" for such variables will get rid of self-selection bias and tell you whether there is hiring prejudice.

The only way to answer the question is with an experiment. And the experiment has been done; we know the answer. The economists Rand Ghayad and William Dickens sent out 4,800 fictitious applications for six hundred job openings.[14] Even when applications were identical except for alleged length of unemployment, the short-term unemployed were twice as likely to get an interview as the long-term unemployed. In fact, the short-term unemployed were more likely to get an interview for a job for which they were not very well qualified than were long-term unemployed people who were more fully qualified!

There are questions that can only be handled by an experiment, but which some scientists nonetheless feel are better answered by multiple regression analysis.

Many experimental studies have shown that African American job applicants with black-sounding names (D'André, Lakaisha) are less likely to get interviews than identical applicants with names that don't sound black (Donald, Linda). Applicants with white-sounding names have as much as a 50 percent greater likelihood of being granted an interview than applicants with black-sounding names.[15] Having a white-sounding name rather than a black-sounding name is worth as much as eight years of job experience. Being dubious about whether black names actually produce worse economic outcomes, the highly respected economists Roland Fryer and Steven Levitt conducted a multiple regression study examining the relationship between having a black-sounding name and various economic outcomes.[16] The population they studied consisted of black women born to non-Hispanic blacks in California who had remained in the state as adults. The dependent variables were not job-hunting success, or income, or occupational status, but indirect measures of life outcomes such as average income in the woman's zip code and whether the woman had private health insurance. The investigators report that the latter variable is "the best measure we have with respect to the quality of her current employment."[17] (The best that the investigators had, that is. It's really a rather crude measure of occupational attainment.)

Fryer and Levitt found that women with black-sounding names fare substantially worse on their indicators of occupational success than women with white-sounding names, as we would expect based on the experimental studies. But the relationship between type of name and the outcome variables vanished when they controlled for variables such as percent of black babies in the hospital of the woman's birth, percent of black babies in the county of her birth, whether the mother was born in California, mother's age at time of birth, father's age at time of birth, months of prenatal care, whether the woman was born in a county hospital, her birth weight, her total number of children, and whether she was a single mother.

The authors were aware of the problems with this kind of analysis. They acknowledge that "the clear weakness of this empirical approach is that if unobserved characteristics of the woman are correlated both with life outcomes and her name, our estimates will be biased."[18] Indeed.

Nevertheless, the authors go on to state that there is no connection between how black-sounding one's name is and life outcomes after they control for other factors. "We find . . . no negative relationship between having a distinctively Black name and later life outcomes after controlling for a child's circumstances at birth."[19] A very large number of variables, many of which would be more predictive of occupational success than the ones Fryer and Levitt examined, would have to have been assessed in order to justify that conclusion. (And when very large numbers of control variables are examined, many of them having stronger relationships with the dependent variable than the correlation of most interest, conclusions become shaky.)

Fryer and Levitt imply that parents can give their child a black-sounding name without worrying that there might be negative effects on occupational success. That seems extraordinarily unlikely in light of the experimental studies.

A recent study by Katherine Milkman and her colleagues shows that a black-sounding name can surely disadvantage candidates for graduate school.[20] Thousands of professors were sent an e-mail message allegedly from a prospective graduate student requesting a meeting a week hence to discuss research opportunities. A male student with a white-sounding name was 12 percent more likely to be granted the interview than a male student with a black-sounding name. Such a difference could have real consequences. Getting one's first choice for graduate advisor versus not can mean a distinguished career versus a less distinguished one.

Why might Fryer and Levitt have been willing to assume that an MRA study could be powerful and accurate enough to cast doubt on the implications of the experimental studies? I suspect it's because of what the French call *déformation professionelle*—the tendency to adopt the tools and point of view of people who share one's profession. For many of the types of research that economists do, MRA is the only available option. Economists can't manipulate interest rates set by the Federal Reserve. If you want to find out whether austerity or pump-priming was more helpful to a country's economy during the Great Recession, you can correlate degree of austerity with strength of recovery, but you can't randomly assign countries to an austerity condition.

Economists are taught MRA as their main statistical tool. But they are not taught to be nearly as critical of it as they need to be. Levitt, in a book cowritten with the journalist Stephen Dubner,[21] reported on an analysis of data collected by the U.S. Department of Education called the Early Childhood Longitudinal Study. The academic achievement of students from kindergarten to fifth grade was examined, along with dozens of other variables, such as parental income and education, how many books were in the child's home, how much the child was read to, whether the child was adopted, and so forth. Levitt reports on the MRA-based conclusions about the relationship between a host of these variables and academic achievement. He concludes that, net of many variables including number of books in the home, "reading books doesn't have an impact on early childhood test scores."[22] MRA simply isn't up to the job of establishing that reading to children is unimportant for their intellectual development. Only an experiment could do that. Levitt had an additional finding indicating that, net of many variables including reading books to children, having books in the home has an important effect on test scores. Thus, owning lots of books makes children smarter, but reading those books to them doesn't. Levitt's faith in MRA is such that he actually attempts to give a causal explanation for this state of affairs.

A much more important error is Levitt's assertion that family environment has relatively little effect on children's intellectual skills. He bases this conclusion on studies of adoptive children. "Studies have shown that a child's academic abilities are far more influenced by the IQs of his biological parents than the IQs of his adoptive parents."[23] But correlations are the wrong data to look at to reach an estimate of the importance of family environment. We need to look instead at the results of the natural experiment of adoption of a child versus leaving the child with the birth parents, who typically are of much lower socioeconomic status. The environments created by adoptive parents are substantially more favorable in many respects than those of families in general. And, in fact, the school performance of adopted children is half a standard deviation higher than that of siblings who were not adopted; the IQs of adopted children are more than a standard deviation higher than nonadopted siblings. And the higher the social class of the adoptive parents (and therefore the more favorable the intellectual environment on average), the higher the IQ of the adopted child. The effects of family environment on intellectual skills are actually very great.[24]

In Levitt's defense, he didn't come up with his mistaken conclusions about the effect of adoptive environments on his own. Behavioral scientists and geneticists have been using the correlational data to reach wrong conclusions about environmental effects on intellectual ability for decades.

Some eminent economists don't seem to recognize the value of experiments at all. The economist Jeffrey Sachs started an extremely ambitious program of health, agricultural, and educational interventions in a small number of African villages, with the intent of improving quality of life. The program's cost is very high relative to alternatives, and it has been severely criticized by other development experts.[25]

Though some of Sachs's villages improved their residents' conditions, similar villages in Africa improved more without his intervention. Sachs could have ended the criticism by randomly assigning similar villages to his treatment condition versus no treatment and showing that his villages made better progress than those control villages. Sachs has refused to conduct this experiment on what he described as "ethical grounds."[26] What's unethical is *not* to conduct experiments when they're feasible. Sachs spent a great deal of other people's money, but we have no idea whether that money improved people's lives more than alternative, possibly less expensive programs would have.

As it happens, though, increasing numbers of economists are conducting social psychology–type randomized control experiments. One recent example is a particularly impressive series of experiments conducted by the economist Sendhil Mullainathan and the psychologist Eldar Shafir showing that resource scarcity can have dire consequences for the cognitive functioning of everyone from farmers to CEOs.[27] If you ask people to imagine how they would rejigger their budgets if they suddenly were confronted with the need for an auto repair costing several thousand dollars and then give them an IQ test, you will find that the IQ of poor people takes a big beating. Well-off people's IQ is unaffected by this thought exercise. (And contemplation of an auto repair of a few hundred dollars doesn't impair the test performance of either poor people or well-off people.)

The economist Raj Chetty is a leader in pushing economists toward finding natural experiments that test economic hypotheses. Does teacher quality really matter in the long run? We can estimate how much difference it makes to have a highly competent teacher versus a much less competent one by looking at the average performance of a given class of students before and just after the entry of a high-value teacher (or after the exit of such a teacher).[28] For example, each cohort of third-grade students in a given school may be getting about the same mediocre scores on achievement tests year after year until the entry of a teacher with a good track record. (Perhaps the previous teacher has left because of an illness.) If the performance of the third-year classes shows a jump that is sustained as long as that teacher is in place, we can look at the effect of that increase on subsequent academic achievement, college attendance, and adult income. And the effects of teacher competence on all these variables are marked. Such studies count as near-experiments because classes prior to the entry of a given teacher are essentially the control group for classes following the entry of the new teacher. Assignment to condition falls short of random, but when teachers' assignments are apparently fortuitous, we have a reasonably good natural experiment.

And some of the most important experiments by economists are those on educational interventions carried out by Roland Fryer. He has conducted extremely valuable educational experiments, showing, for example, that financial incentives have little effect on the academic achievement of minority students.[29] They also have little effect on teacher performance—except when loss aversion is triggered by giving the incentives at the beginning of the year with the understanding that failure to improve student achievement will result in losing the incentive.[30] This finding, incidentally, is a lovely example of the greater effect of potential loss than of potential gain, discussed in Chapter 5. Fryer has also played a role in the very successful Harlem Children's Zone experiments, which yielded large increases in academic achievement for African American students.

My Tribe, Too

I'm afraid I now have to admit that psychologists can be as guilty of misusing MRA as other behavioral scientists.

It's common to report the following kind of finding. Employees of companies with generous parental leave benefits are more satisfied with their jobs than employees of companies that don't provide that benefit. This correlation is then buttressed with an MRA showing that the better the leave policy, the more satisfied employees are with their jobs, and that this is still true when you "control for" size of company, employee salary, ratings of how pleasant coworkers are, ratings of how much the immediate superior is liked, and so forth. There are three problems with this kind of analysis. First, a limited number of variables will have been measured, and if one or more have been poorly measured, or if there are other variables not examined by investigators that are correlated both with generosity of parental leave policy and job satisfaction, it may be those associations that account for job satisfaction, not leave policy. Second, it really makes no sense to pull parental leave policy out of the total picture of the employee's experience with a company. Generosity of the company in that respect is likely to be bound up with all kinds of other positive qualities of the company. Pulling that one thread out of the complicated ball of relationships among variables, then attempting to "control" for a few out of many variables in that ball, is not likely to protect us from mistakes. Third, these kinds of analyses leave us particularly vulnerable to the halo effect problem discussed in Chapter 3. People who like their jobs also find the restrooms to be cleaner, the employees to be better looking, and the commutes to be less tedious than people who don't much like their jobs. Love is blind, and liking's vision isn't all that much better.

These problems are perhaps easier to see in the case of personality research. It can make little sense to single out one aspect of a person's character and assume it's not strongly related to—enmeshed with—other aspects of the person's character. It's common for psychologists to report findings such as "Self-esteem is correlated with academic performance, controlling for extroversion, measures of self-control, depressive tendencies," and so on. But low self-esteem, and other undesirable states, such as depression, are generally found to be correlated: when you're down you don't think too highly of yourself, and when you think poorly of yourself that's likely going to pull your mood down. Looking at self-esteem as if it could be pulled out of the relationship to depression is just an artificial thing to do. It's not plausible that there are many people who could say, "I think I'm terrific, too bad I'm so depressed I can hardly see straight," or "I've never been happier, too bad I'm such a jerk." It's possible, maybe, but the odd ring of such sentences reflects the fact that self-esteem and depression are normally tangled up together. They're not a composite whose elements can be separated.

Many of my fellow psychologists are going to be distressed by my bottom line here: such questions as whether academic success is affected by self-esteem, controlling for depression, or whether the popularity of fraternity brothers is affected by extroversion, controlling for neuroticism, or whether the number of hugs a person receives per day confers resistance to infection, controlling for age, educational attainment, frequency of social interaction, and a dozen other variables, are not answerable by MRA. What nature hath joined together, multiple regression analysis cannot put asunder.

No Correlation Doesn't Mean No Causation

Correlation doesn't prove causation. But the problem with correlational studies is worse than that. *Lack of correlation doesn't prove lack of causation*—and this mistake is made possibly as often as the converse error.

Does diversity training improve rates of hiring women and minorities? One study examined this question by quizzing human resource managers at seven hundred U.S. organizations about whether they had diversity training programs and by checking on the firms' minority hiring rates filed with the Equal Employment Opportunity Commission.[31] As it happens, having diversity training programs was unrelated to "the share of white women, black women, and black men in management." The authors concluded that diversity training did not affect minority hiring.

But just a moment. Having diversity training versus not having it is a self-selected variable. Some corporations that hire diversity trainers may be less interested in hiring women and minorities than corporations that find more effective ways to increase hiring. In fact, they may simply be using such programs as protective cover for their real hiring policies. Corporations that don't have diversity training may be effective in hiring minorities by techniques such as setting up diversity task forces or, as the U.S. military does, making successful minority advancement a part of the evaluation of superior officers. Proving that diversity training does or doesn't work will require randomized experiments. We have to fight the reflexive conclusion that A can't exert a causal influence on B because there is no correlation between the two.

Discrimination: Look at the Statistics
or Bug the Conference Room?

While we're on the topic of discrimination, let me point out that you can't prove whether discrimination is going on in an organization—or a society—by statistics. You often read about "glass ceilings" for women in a given field or about disproportionate school suspensions of boys or minorities. The intimation—often the direct accusation—is that discrimination is at work. But numbers alone won't tell the story. We don't know that as many women as men have the qualifications or desire to be partners in law firms or high-level executives in corporations. And we have some pretty good reasons to believe that girls and boys are not equally likely to engage in behavior warranting suspension from school.

Not so long ago, it was common to attribute women's lower representation in graduate school and faculty rosters to discrimination. And there certainly was discrimination. I know; I was there. I was privy to the conversations the men had about admitting women to grad school or hiring them onto faculties. "Go after the guy; women are too likely to drop out." Bugged conversations would have proved what raw statistics, comparing percentage of men and women hired, could not.

But nowadays 60 percent of college graduates are women, and they constitute a majority of law and medical students as well as graduate students in the humanities, social sciences, and biological sciences. At the University of Michigan, where I teach, two-thirds of the assistant professors hired are women (and they get tenure at the same rate as men).

Do these statistics prove discrimination against men? They do not. And I can assure you that bugged conversations—at least at my school— would not support the discrimination idea either. On the contrary, we are so frequently confronted with the prospect of admitting huge majorities of women into our graduate program that we contemplate relaxing admission standards for men, though we've never carried it out in a conscious way, of that I'm sure.

The statistics on postgraduate education have not stopped some people from claiming there is still discrimination against women in the physical sciences. One book I read recently claimed that women were "locked out" of physics. In the absence of evidence other than the purely statistical kind, there can be no justification for that assertion.

But we don't have to resort to bugging conference rooms to establish that discrimination exists. Experiments can do the job. Do car sales people cite higher prices for women and minorities than for white men? Send a white man, a woman, and a minority group member to Mammoth Motors and see what price each gets quoted. The study has been done, and in fact the white man is quoted the lowest price.[32]

Do good-looking people get better breaks in life? Many studies show that to be the case. Clip a picture of an alleged delinquent to a file and see what sentence is recommended by undergraduate "judges." If the kid is good-looking, he's seen as likely to become a better citizen in the future and is given a relatively light sentence. If the kid is ugly, throw the book at him.[33]

"Life is unfair," as John F. Kennedy said, and experiments are the best instrument we have for revealing just how much more unfair it is for some groups of people than for others.

Summing Up

Multiple regression analysis (MRA) examines the association between an independent variable and a dependent variable, controlling for the association between the independent variable and other variables, as well as the association of those other variables with the dependent variable. The method can tell us about causality only if all possible causal influences have been identified and measured reliably and validly. In practice, these conditions are rarely met.

The fundamental problem with MRA, as with all correlational methods, is self-selection. The investigator doesn't choose the value for the independent variable for each subject (or case). This means that any number of variables correlated with the independent variable of interest have been dragged along with it. In most cases, we will fail to identify all these variables. In the case of behavioral research, it's normally certain that we can't be confident that we've identified all the plausibly relevant variables.

Despite the above facts, MRA has many uses. Sometimes it's impossible to manipulate the independent variable. You can't change someone's age. Even when we have an experiment, it adds to our confidence to know that the experimentally demonstrated relationship holds in a natural ecology. And MRA is in general vastly cheaper than experiments, and it can identify relationships that would be important to examine experimentally.

When a competently conducted experiment tells you one thing about a given relationship and MRA tells you another, you normally must believe the experiment. Of course, a badly conducted experiment tells you no more than MRA, sometimes less.

A basic problem with MRA is that it typically assumes that the independent variables can be regarded as building blocks, with each variable taken by itself being logically independent of all the others. This is usually not the case, at least for behavioral data. Self-esteem and depression are intrinsically bound up with each other. It's entirely artificial to ask whether one of those variables has an effect on a dependent variable independent of the effects of the other variable.

Just as correlation doesn't prove causation, absence of correlation fails to prove absence of causation. False-negative findings can occur using MRA just as false-positive findings do—because of the hidden web of causation that we've failed to identify.

12. Don't Ask, Can't Tell

How many questionnaire and survey results about people's beliefs, values, or behavior will you read during your lifetime in newspapers, magazines, and business reports? Thousands, surely. You may even create some of these surveys yourself in order to get information that is important for your business, school, or charitable organization.

Most of us tend to read survey results rather uncritically. "Hmm, dear, I see in the *Times* that 56 percent of Americans favor tax increases for creating more national parks." Ditto for questions we create ourselves and the answers our respondents give us.

So far, all the methods I've discussed are applicable to pretty much everything—animal, vegetable, or mineral. We can do A/B testing on rats, learn from natural experiments about factors influencing corn yields, and do multiple regression studies of factors associated with water purity. Now I'd like to look at methodological difficulties in measuring specifically human variables. Unlike rats, corn, or water, people can tell you in verbal form (oral or written) about their attitudes, emotions, needs, goals, and behavior. And they can tell you what the causal influences on these variables are. In this chapter you'll see just how misleading such reports can be, which won't be surprising given what you read in Part I about our limited accessibility to the factors that influence our behavior. The chapter will show you how a variety of behavioral measures can provide much more trustworthy answers to questions about people's attributes and states than their verbal reports can.

You'll also get some tips about experiments you can do on yourself to learn what kinds of things influence your attitudes, behaviors, and physical and emotional health. Correlational evidence about yourself can be just as misleading as correlational evidence about anything else. Experiments on yourself can produce evidence that is accurate and compelling.

Constructing Attitudes on the Fly

The following examples may make you pause before trusting self-reported verbal answers, and may help you consider how best to get useful information about people's attitudes and beliefs. The examples may also increase your doubts about people's explanations of the causal influences on their judgments and behavior.

Q. Suppose I ask you about three positive events in your life and then ask you about your life satisfaction; *or* I ask you about three negative events and then about your life satisfaction. In which case do you report greater life satisfaction?

A. Whatever you guessed about the effect of asking about positive versus negative events, I'm sorry to tell you your answer is wrong. It all depends on whether those events I asked you about were in the recent past or happened five or so years ago. Your life seems worse if you've just contemplated some lousy things that have happened lately than if you've contemplated some good things that have happened lately.[1] No surprise there. But the reverse is true if you contemplate events from five years ago. Your life seems good compared to the bad things that happened in your past. And your life seems not so great compared to the wonderful things that used to happen. (This helps to explain the otherwise puzzling fact that for members of the Greatest Generation, life satisfaction is greater the worse their experiences during the Depression had been.)[2]

Q. Your cousin from Omaha calls you and asks you how things are going. Is your answer influenced by whether it's sunny and warm versus cloudy and cold where you are?

A. Turns out that it depends. If the weather is nice, you're more likely to say things are going well than if the weather is lousy. Well, of course. *But* . . . if your cousin inquires first about the weather in your city today and then asks you how things are going, there is no effect of the weather on your report of how things are going.[3] Why? Psychologists say that when prompted to think about the weather, we *discount* some of our mood as being weather-related and add or subtract happiness points accordingly. In effect: "Life seems to be going pretty well, but probably part of the reason I feel that way is that it's seventy degrees and sunny out, so I guess things are just so-so."

Q. What do you suppose is the correlation between satisfaction with one's marriage and satisfaction with one's life as a whole?

A. This seems like a fairly easy thing to examine. We can ask people about their satisfaction with their lives and then ask them about their satisfaction with their marriages. The higher the correlation between the two, the greater we might assume the impact of marriage satisfaction on life satisfaction to be. That correlation has been examined.[4] The correlation is .32, indicating a modestly important effect of marriage satisfaction on life satisfaction as a whole. But suppose we reverse the question order and ask people how satisfied they are with their marriages before we ask how satisfied they are with their lives. Now the correlation is .67, indicating a major effect of marriage quality on life quality. So whether Joe tells us that life is good or just all right depends—and depends *heavily*—on whether you just asked him how good his marriage is. This phenomenon, and many others discussed in this chapter, shows the effects of verbal priming of the type discussed in Chapter 1 on people's reports about their attitudes. Other phenomena show the influence of context of the kind discussed in Chapter 2 on reports about attitudes.

The likely reason question order is so important is that asking first about marriage makes it highly salient, so it heavily influences the respondent's feelings about life overall. If you don't ask first about marriage, the respondent considers a broader range of things, and that wider set of influences figures into the assessment of life satisfaction. So just how important, really, is marriage quality for life quality? There can be no answer to that question. At any rate, not by asking questions of this kind. If the apparent importance of marriage quality for life quality is so malleable, then we've learned little about the reality.

But the truth is, the answer to just about every question concerning attitudes and behavior can be pushed around—often by things that seem utterly fortuitous or silly.

Suppose I ask you how favorable you are toward politicians. But before I do, I point out that the average rating of politicians given by other people is 5 on a scale of 1–6, with higher numbers being more favorable. Or I point out that the average rating of politicians is 2 on that scale. You will rate politicians higher in the first case than in the second. Some of that is due to sheer conformity. You don't want to seem an oddball. But more interesting, announcing others' ratings tacitly changes not just your judgment of politicians but your assumptions about the kind of politicians I'm asking about.[5] If I tell you most people have a high opinion of politicians, I've implied that by "politicians" I mean statesmen on the order of Churchill or Roosevelt. If I tell you that most people have a low opinion of politicians, I have tacitly implied that by "politicians" I mean hacks and chiselers. I've literally changed what it is you're making your judgment about.

What percent of Americans are in favor of the death penalty? In the abstract, a majority. For any given case, a minority. The more details we present about the crime, the criminal, and the circumstances, the less inclined respondents are to be willing to execute the perpetrator.[6] Remarkably, that's true even for the most heinous of crimes, such as a criminal who rapes women and then kills them. The more details you give about the perpetrator's character and life history, the more reluctant people are to favor the death penalty. This is true even when that information is overwhelmingly negative.

What percent of Americans support abortion? Here I close the blinds and ask, sotto voce, "What do you want it to be?" According to a 2009 Gallup poll, 42 percent of Americans say they are "pro-choice" as opposed to "pro-life."[7] So 42 percent of Americans support abortion. But according to another Gallup poll the same year, 23 percent believe that abortion should be legal in all circumstances and 53 percent believe that abortion should be legal under certain circumstances.[8] So 76 percent of Americans support abortion. I have no doubt that we could get that percentage higher still if we asked whether the respondent favors abortion in the case of rape, in the case of incest, or in order to save the life of the mother. If the respondent replies yes to any of those questions, we can record the respondent as favoring abortion. So whether less than half the population supports abortion or a heavy majority supports it is entirely a matter of question wording.

A host of studies by psychologists show that people don't carry all their attitudes around with them in a mental file drawer. "How do I feel about abortion? Hmm. I'll check. Let's see: abortion, attitudes toward. Ah yes, here I have it. I'm moderately opposed."

Instead, many attitudes are extremely *context dependent* and constructed on the fly. Change the context and you change the expressed attitude. Sadly, even trivial-seeming circumstances such as question wording, the type and number of answer categories used, and the nature of the preceding questions are among the contextual factors that can profoundly affect people's reports of their opinions. Even reports about attitudes of high personal or social importance can be quite mutable.

What Makes You Happy?

Verbal reports about attitudes are susceptible to a host of other methodological problems. People lie about some things. Sex. Money. People want to look good in their own eyes and in the eyes of others. This *social desirability bias* often causes people to accentuate the positive and eliminate the negative. But lies and trying to look good are really the least of our problems in finding out the truth about people's attitudes and behavior, and why they believe what they believe and do what they do.

At least we're pretty good at knowing what makes us happy or unhappy. Or are we?

Rank the following factors in order of the degree to which they seem to influence your mood on a given day. Let's see how accurately you can assess what causes your mood to fluctuate. Rate the importance of the following items on a scale of 1 (very little) to 5 (a great deal).

1. How well your work went
2. Amount of sleep you got the preceding night
3. How good your health is
4. How good the weather is
5. Whether you had any sexual activity

6. Day of the week
7. If you are a woman—stage of the menstrual cycle

No matter what you said, there's no reason to believe it's accurate. At any rate, we know that's the case for Harvard women.[9] Psychologists asked students to report at the end of each day, for two months, the quality of their mood. Respondents also reported the day of the week, the amount of sleep they had the night before, what their health status was, whether they had had any sexual activity, what stage of the menstrual cycle they were in, and so forth. At the end of the two months, participants were asked how each of the factors tended to affect their mood.

The participants' answers to these questions made it possible to find out two things: (1) how much participants thought each factor affected their mood, and (2) how well each factor actually predicted their mood. Did these self-reports reflect the actual correlations between reported factors and reported moods?

As it turned out, participants were not accurate at all. There was zero correlation between a factor's actual effect on mood (based on the daily ratings) and participants' beliefs about the degree to which variations in the factor influenced variations in mood. Literally no correspondence at all. If the woman said day of the week was very important, the actual association between day of the week and mood was as likely to be low as high. If the woman said sexual activity was not very important, the actual correlation between sexual activity and mood was as likely to be high as low.

There was an even more embarrassing finding. (Embarrassing to the participants, but also to everybody else, since there's no reason to assume Harvard women are uniquely lacking in insight into the causes of their mood.) Jane's self-reports about the relative influence of the factors affecting her mood were no more accurate than her guesses about the effects of those factors on a typical Harvard woman's mood. In fact, her guesses about the typical student were pretty much the same as her guesses about herself.

Clearly, we have theories about what affects our moods. (Goodness knows where we get them all.) When asked how various things affect our mood, we consult these theories. We're unable to access the facts, even though it feels as though we can.

I'm tempted to say we don't know what makes us happy. That goes too far, of course. What we can say is that our beliefs about the relative importance of different events affecting our well-being are poorly calibrated with their actual importance. Of course there's nothing unique about the factors affecting mood. As you read in Chapter 8 on correlations, detecting correlations of any kind is not one of our strong suits.

The lesson of the Harvard study is a general one. Psychologists find that our reports about the causes of our emotions, attitudes, and behavior can be quite untrustworthy, as was first shown in Part I.

The Relativity of Attitudes and Beliefs

> First man: "How's your wife?"
> Second man: "Compared to what?"   —Old vaudeville routine

Test the validity of your opinions about ethnic and national differences by answering the following questions:

Who values being able to choose personal goals more: Chinese or Americans?
Who are more conscientious: Japanese or Italians?
Who are more agreeable: Israelis or Argentineans?
Who are more extroverted: Austrians or Brazilians?

I'm betting you didn't guess that Chinese value choosing their own goals more than Americans,[10] or that the Italians are more conscientious than the Japanese, the Israelis more agreeable than the Argentineans, or the Austrians more extroverted than the Brazilians.[11]

How do we know these differences exist? People from those countries tell us so themselves.

How could people's beliefs about their values and personalities differ so much from popular opinion? (And for that matter, from the opinions of academic experts who are highly familiar with each of the cultural pairs above.)

People's answers about their own values, traits, and attitudes are susceptible to a large number of *artifacts*. (The word "artifact" has two dimly related meanings. In archaeology, the word refers to an object created by humans, for example, a piece of pottery. In scientific methodology, the word refers to a finding that is erroneous due to some unintended measurement error, often due to intrusive human action.)

In the case of the cultural comparisons above, the discrepancy between people's self-reports about their characteristics and our beliefs about the characteristics of people of their nationality is due to the *reference group effect*.[12] When you ask me about my values, my personality, or my attitudes, I base my answer in part on a tacit comparison with some group that is salient to me, for example because I'm a member of it. So an American, asked how important it is to be able to choose her own goals, implicitly compares herself to other Americans, and perhaps to other Jewish Americans, and perhaps to other Jewish American females in her college. So compared to other Americans (or Jews, or Jewish females, or Jewish females at Ohio State), choosing her own goals doesn't seem like all that big a deal to her. The Chinese respondent is comparing himself to other Chinese, or other Chinese males, or other Chinese males at Beijing Normal University—and it may seem to him that he cares more about choosing his own goals than do most people in his reference group.

One reason we know that tacit comparison with a reference group is a big factor in producing these self-reports (Austrians more extroverted than Brazilians, etc.) is that they disappear when you make the reference group explicit. European Americans at Berkeley rate themselves as more conscientious than do Asian Americans at Berkeley, but not when you have both groups compare themselves to the explicit reference group of "typical Asian American Berkeley students."[13]

Other things being equal, people in most cultures believe they are superior to most others in their group. This *self-enhancement bias* is sometimes known as the Lake Wobegon effect, after Garrison Keillor's mythical town where "all the children are above average." Seventy percent of American college students rate themselves as above average in leadership ability, and only 2 percent rate themselves below average.[14] Virtually everyone self-rates as above average in "ability to get along with others." In fact, 60 percent say they are in the top 10 percent and 25 percent say they are in the top 1 percent!

Degree of self-enhancement bias differs substantially across cultures and across subgroups within a given culture. No one seems to top Americans in this respect, whereas East Asians often show a contrary effect, namely a *modesty bias*.[15] So any self-assertions concerning issues having a value component (leadership, ability to get along with others) will find Westerners rating themselves higher than East Asians do. Americans will rate themselves as better leaders than Koreans do and Italians will rate themselves as more conscientious than Japanese do.

Many other artifacts find their way into self-reports. These include what's called *acquiescence response set* or *agreement response bias*. This is the tendency to say yes to everything. As you might expect, yea-saying is more common among polite East Asians and Latin Americans than it is among frank Europeans and European Americans. There are also individual differences within a culture in tendency to agree. Fortunately, there's a way to counteract this: investigators can *counterbalance* response categories so that half the time respondents get a high score on some dimension—extroversion versus introversion, for example—by agreeing with a statement and half the time by disagreeing with a statement. ("I like to go to large parties" versus "I don't like to go to large parties.") This cancels out any bias to agree with statements in general. The counterbalancing correction is well known to all social scientists but is surprisingly often neglected by them.

Talking the Talk Versus Walking the Walk

But is there a better way to compare people, groups, and whole cultures than just by asking them? You bet there is. Behavioral measures, especially those taken when people don't realize they're being observed, are much less susceptible to artifacts of all kinds.

Rather than ask people how conscientious they are, you can measure how conscientious they are by examining their grades (or better, their grades controlling for their cognitive ability scores), the neatness of their rooms, how likely they are to be on time for an appointment or a class, and so on. We can also examine the conscientiousness of whole cultures by measuring such proxies for conscientiousness as speed of postal delivery, accuracy of clocks, on-time record of trains and buses, longevity, and number of questions people answer on a lengthy and boring questionnaire. (Incidentally, the correlation between the math scores of different nations and the number of tedious questions they answer on an interminable questionnaire is extremely high.)

Remarkably, it turns out that when we examine behavior to find out how conscientious people of different countries are, we find that the less conscientious a nation is as measured by behavioral indices, the more conscientious its citizens are as measured by self-report![16]

When it comes to the measurement of virtually any psychological variable, I follow the maxim that you should trust behavior (including physiological behavior such as heart rate, cortisol output, and the activity of different brain regions) more than responses to concrete *scenarios* (descriptions of situations followed by measures of expected or preferred outcomes or behaviors by the self or others). In turn, you should trust scenario responses more than verbal reports about beliefs, attitudes, values, or traits.

I wouldn't wish to have you doubt every verbal report you see in the media or doubt your own ability to construct a questionnaire of any kind. If you want to find out whether your employees would rather have the picnic on a Saturday or a Sunday, you don't have to worry much about their answers being valid.

But even for expressions of preference, you can't necessarily trust self-reports. As Steve Jobs said, "It's not the customers' job to know what they want." Henry Ford remarked that if he had asked people what they wanted in the way of transportation, they would have said "faster horses." And Realtors have an expression: "Buyers are liars." The client who assures you she must have a ranch house falls in love with a 1920s Tudor. The client who pines after a modern steel and glass edifice ends up with a faux adobe house.

Finding out people's preferences is a tricky matter for businesses. Even the best thought-out *focus group* can come a cropper. Henry's successors at Ford Motor Company had a fondness for focus groups, in which a group of people are quizzed by corporate representatives and by each other; the organizers use the expressed preferences to establish what new goods or services would be likely to succeed. Automotive legend has it that in the mid-1950s, Ford had the idea of removing the center post from a four-door sedan to see whether its sporty appearance would appeal to buyers. The people they gathered for the focus groups thought the idea was a bad one: "Why, it hasn't got a center post." "It looks weird."

"I don't think it would be safe." General Motors skipped the focus groups and went straight into production with a center-post-free Oldsmobile, calling it a four-door hardtop convertible. It was a huge success. The hardtop experience apparently didn't cause Ford to rethink how much attention they should pay to focus groups. The company doubled down on them in making their decision to market the 1950s Edsel—the very icon of product failure.[17]

The take-home lesson of this section: whenever possible don't listen too much to people talk the talk, watch them walk the walk.

More generally, the chapters in this section constitute a sermon about the need to get the best possible measures of any variable that we care about and find the best possible means to test how it's related to other variables. In the great chain of investigation strategies, true experiments beat natural experiments, which beat correlational studies (including multiple regression analyses), which, any day, beat assumptions and Man Who statistics. Failure to use the best available scientific methodology can have big costs—for individuals, for institutions, and for nations.

Experiments on Yourself

As shown by the Harvard study of women asked to assess factors influencing their moods, we are in as much trouble detecting correlations in our own lives as in other areas. Fortunately, we can do experiments with ourselves as the subject and get better information about what makes us tick.

What factors make it difficult to fall asleep? Is coffee in the morning helpful for efficiency during the day? Do you get more useful work done in the afternoon if you take a catnap after lunch? Are you more effective if you skip lunch? Does yoga improve well-being? Does the Buddhist practice of "loving-kindness"—visualizing smiling at others, reflecting on their positive qualities and their acts of generosity, and repeating the words "loving-kindness"—bring you peace and relieve you of anger toward others?

A problem with experiments on the self is that you're dealing with an N of 1. An advantage, however, is that experiments on the self automatically have a within, before/after design, which can improve accuracy because of the reduction in error variance. You can also keep confounding variables to a minimum. If you're looking to discover the effect of some factor on you, try to keep everything else constant across the study period when you're comparing presence of the factor versus absence of the factor. That way you can have a fairly good experiment. Don't take up yoga at the same time as you move from one house to another or break up with your boyfriend. Arrange to start yoga when a proper before/after design is possible. Monitor your physical and emotional well-being, the quality of your relations with others, and your effectiveness at work for a few weeks before taking up yoga, and use the same measures for a few weeks after taking it up. Simple three-point scales provide adequate measures of these things. At the end of the day rate your well-being: (1) not great, (2) okay, (3) very good. Get the mean on each variable for the days before taking up yoga and for the days after. (And hope nothing big happens in your life to muddy the waters.)

Often you can do better than the before/after study. You can take advantage of random assignment to condition. If you try to figure out whether coffee in the morning improves your efficiency, don't just drink coffee haphazardly. If you do, any number of confounding variables can distort the test results. If you drink coffee only when you feel particularly groggy in the morning, or only on a day when you have to be at the top of your form at work, your data are going to be a mess, and any lesson you'll think you've learned will likely be off the mark. Literally flip a coin as you walk into the kitchen—heads you have coffee, tails you don't. Then keep track—in writing!—of your efficiency during the day. Use a three-point scale: not very efficient, fairly efficient, very efficient. Then after a couple of weeks do a tally. Calculate the mean effectiveness on days with and without coffee.

The same experimental procedure works for any number of things that are candidates for influencing your well-being or effectiveness. And don't kid yourself that you can figure these things out without being systematic about random assignment to condition and rigorously keeping track with decent measures of outcomes.

It's eminently worth doing experiments like this because there are actually big individual differences in things such as the effects of coffee, the degree of benefit from both endurance training and weight training, and whether peak work efficiency is in the morning, afternoon, or evening. What works for Jill or Joe may not work for you.

Summing Up

Verbal reports are susceptible to a huge range of distortions and errors. We have no file drawer in our heads out of which to pull attitudes. Attitude reports are influenced by question wording, by previously asked questions, by "priming" with incidental situational stimuli present at the time the question is asked. Attitudes, in other words, are often constructed on the fly and subject to any number of extraneous influences.

Answers to questions about attitudes are frequently based on tacit comparison with some reference group. If you ask me how conscientious I am, I will tell you how conscientious I am compared to other (absent-minded) professors, my wife, or members of some group who happen to be salient because they were around when you asked me the question.

Reports about the causes of our behavior, as you learned in Chapter 3 and were reminded of in this chapter, are susceptible to a host of errors and incidental influences. They're frequently best regarded as readouts of theory, innocent of any "facts" uncovered by introspection.

Actions speak louder than words. Behavior is a better guide to understanding people's attitudes and personalities than are verbal responses.

Conduct experiments on yourself. The same methodologies that psychologists use to study people can be used to study yourself. Casual observation can mislead about what kinds of things influence a given outcome. Deliberate manipulation of something, with condition decided upon randomly, plus systematic recording, can tell you things about yourself with an accuracy unobtainable by simply living your life and casually observing its circumstances.

PART V

THINKING, STRAIGHT AND CURVED

People have discovered many different ways to reduce the likelihood of making an error in reasoning. One way is to obey the rules of *formal logic*—rules for reasoning that can be described in purely abstract terms without making any contact at all with real-world facts. If the structure of your argument can be mapped directly onto one of the valid forms of argument that logic specifies, you're guaranteed a *deductively valid conclusion*. Whether your conclusion is true is a different matter entirely and depends on the truth of your *premises*—the statements that precede your conclusion. Formal logic is a type of *deductive reasoning*—"top-down" argument forms that produce conclusions that follow necessarily from the premises on which they are based.

Two kinds of formal logic have received a great deal of attention historically. The oldest is the *syllogism*. Syllogisms are used for some kinds of categorical reasoning. For example: All A are B, X is an A, therefore X is a B. (Most famously: All men are mortal, Socrates is a man, therefore Socrates is mortal.) Syllogisms have been around for at least twenty-six hundred years.

Formal logic also includes *propositional logic*, which is somewhat more recent, having first been treated seriously by fourth-century B.C. Greek Stoic philosophers. This kind of logic tells us how to reach valid conclusions from premises, such as with the *logic of the conditional*. For example: If P is the case, then Q is the case. P is the case, therefore Q is the case. (If it snows, the schools will be closed. It snowed. Therefore the schools will be closed.) P is a condition requiring Q, or, differently put, P is a sufficient condition for Q.

In contrast to deductive logic, *inductive reasoning* is a "bottom-up" type of reasoning. Observations are collected that suggest or support some conclusion. One type of inductive reasoning consists of observing facts and reaching a general conclusion about facts of their particular kind. This book is full of different types of inductive reasoning. The scientific method nearly always involves—in fact often is completely dependent on—inductive reasoning of one kind or another. All of the types of inductive reasoning in this book are *inductively valid*, but their conclusions are not deductively valid, merely probable. On the basis of observation and calculation we induce that the mean of the population of some events is X plus or minus Y standard deviations. Or we induce from observing the results of our experiment that A causes B, since we observe that every time A is the case B is also the case; when A is not the case, B is not the case. It's more probable that A causes B if these things are true than if we're missing those observations, but it's not certain that A causes B. For example, something associated with A might be causing B. Inductive conclusions aren't guaranteed to be true even if all the observations they are based on are true, there are many of them, and there are no exceptions. The generalization "all swans are white" is inductively valid but, as it turns out, not true.

Deductive and inductive reasoning schemas essentially regulate inferences. They tell us what kinds of inferences are valid and what kinds are invalid. A very different kind of system of reasoning, also developed about twenty-six hundred years ago in Greece, and developed at the same time in India, is called *dialectical reasoning*. This form of reasoning doesn't so much regulate reasoning as suggest ways to solve problems. Dialectical reasoning includes the Socratic dialogue, which is essentially a conversation or debate between two people trying to reach the truth by stimulating critical thinking, clarifying ideas, and discovering contradictions that may prompt the discussants to develop views that are more coherent and more likely to be correct or useful.

Eighteenth- and nineteenth-century versions of dialectical reasoning, owing primarily to the philosophers Hegel, Kant, and Fichte, center on the process of "thesis" followed by "antithesis" followed by "synthesis"—a proposition followed by a potential contradiction of that proposition, followed by a synthesis that resolves any contradiction.

Other types of reasoning that have been labeled "dialectical" were developed in China, also beginning around twenty-six hundred years ago. Chinese dialectical reasoning deals with a much broader range of issues than Western or Indian versions of dialectical reasoning. The Chinese version suggests ways of dealing with contradiction, conflict, change, and uncertainty. For example, whereas the Hegelian dialectic is "aggressive" in the face of contradiction in the sense that it seeks to obliterate contradictions between propositions in favor of some new proposition, Chinese dialectical reasoning often seeks to find ways in which the conflicting propositions can both be true.

Dialectical reasoning isn't formal or deductive and usually doesn't deal in abstractions. It's concerned with reaching true and useful conclusions rather than valid conclusions. In fact, conclusions based on dialectical reasoning can actually be opposed to those based on formal logic. Relatively recently, psychologists both in the East and in the West have begun to study dialectical reasoning, developing systematic descriptions of prior formulations and proposing new dialectical principles.

Chapter 13 presents two common types of formal reasoning, and Chapter 14 presents an introduction to some forms of dialectical reasoning that I find most interesting and helpful. All of the scientific tools discussed in this book depend to some degree on formal logic. Many of the other tools appeal to dialectical precepts.

13. Logic

Below are four cards. They're randomly chosen from a deck of cards in which every card has a letter on one side and a number on the other. Please indicate which of the cards you would have to turn over in order to find out whether the card obeys this rule: "If a card has a vowel on one side, then it has an even number on the other side." Turn over only those cards that are necessary to establish that the rule is being followed. Commit yourself: if you're reading this on an electronic gadget, highlight your choice in yellow; if you're reading it in hard copy, check your choice with a pencil.

Card 1	Card 2	Card 3	Card 4
N	4	A	3

I must turn over:

a. Card 3 only
b. Cards 1, 2, 3, and 4
c. Cards 3 and 4
d. Cards 1, 3, and 4
e. Cards 1 and 3

We'll return to this problem later in a different context.

•

Critical reasoning texts usually have a heavy dose of formal, deductive logic. This is done more because of ancient pedagogical tradition than because of any evidence of utility for everyday thought. And in fact, there are some reasons to suspect that most of what you read in this chapter on formal logic will be of limited value in solving problems in everyday life.

But there are nevertheless some good reasons to read about formal logic.

1. Formal logic is essential for science and mathematics.
2. The chapter sets up the stark contrast that exists between Western hyperrationality and Eastern dialectical habits of thought. The two systems of thought can both be applied to the same problem, generating different conclusions. The two systems also provide excellent platforms for critiquing each other.
3. An educated person should have some command over basic forms of logical reasoning.
4. Formal logic is interesting, at least to lots of people. (At any rate, in doses about the size of this chapter!)

The story about the Western origins of formal logic goes as follows: Aristotle got tired of hearing lousy arguments in the marketplace and the assembly. So he decided to develop reasoning templates to apply to arguments in order to analyze them for *validity*. An argument is valid if (and only if) its conclusions follow necessarily from its premises. Validity has nothing to do with truth. An argument can be invalid, but its conclusion can be true. An argument is valid if it has the proper structure, but its conclusion can nevertheless be false.

The concept of argument validity is important for many reasons. One reason is that you don't want to allow someone to trick you (or allow you to trick yourself) by conferring plausibility on a conclusion because it follows from some premises—unless those premises are true and the conclusion follows necessarily from the premises. A second reason is that we don't want to allow ourselves to disbelieve conclusions that we happen to dislike if the premises are clearly true and the form of the argument dictates that the conclusion must be true as well. A third reason is that if we have a clear grasp of the concept of validity as opposed to truth, we can assess whether a conclusion follows from its premises by stripping away the meaning from the premises and the conclusion, and thinking purely in terms of abstractions. As and Bs rather than birds and bees. That can show us that a conclusion at least follows from its premises, so even though the conclusion is highly implausible, it's at least not the product of illogical reasoning.

Syllogisms

One of Aristotle's major contributions to formal logic was the syllogism. The creation of syllogisms ballooned into a cottage industry in the Middle Ages, when monks generated dozens of them. From the Middle Ages until the late nineteenth century, philosophers and educators believed that syllogisms provided powerful rules for thought. Consequently they formed a large part of the curriculum in higher education in the West.

The issue of validity arises for syllogisms, which deal with *categorical reasoning*. Some kinds of categorical reasoning involve the *quantifiers* "all," "some," and "none." The simplest syllogisms involve two premises and one conclusion. The simplest of those simple syllogisms, and one that we're not normally at risk of getting wrong, is: All A are B, all B are C, all A are C. The classic here is:

All clerks are human.
All humans are bipedal.

_____

All clerks are bipedal.

That argument is valid because it follows logically from the premises. The conclusion is also true.

All clerks are human.
All humans have feathers.

_____

All clerks have feathers.

That argument is valid, too, though the conclusion is untrue. But the implausibility of the conclusion draws us toward feeling that the argument is also invalid. Substituting As, Bs, and Cs for clerks, humans, and feathers allows us to see the validity of the argument. That may force us to reconsider whether a conclusion is true, which can be useful.

The argument below is invalid, even though its premises and its conclusion are all true (or at least highly plausible).

All people on welfare are poor.
Some poor people are dishonest.
Therefore, some people on welfare are dishonest.

In abstract terms:

All A are B.
Some B are C.
Therefore, some A are C.

The exercise of abstracting the terms is useful because we may have the feeling that a conclusion is true because it seems plausible and because we have true premises that seem to support it logically. Finding that an argument is invalid strips away the feeling that the conclusion is necessarily true and may cause us to doubt it. (The key to recognizing the invalidity of the above argument is to realize that the As are a subset of the Bs.)

Things rapidly become more complicated from here: All A are B, some C are A, some C are B. Valid or not? No A are B, some C are B, no A are C. Valid or not?

You can spin these out till the cows come home. Medieval monks whiled away the weary hours by generating endless numbers of them. But I agree with the philosopher Bertrand Russell, who said that the syllogisms were as sterile as the monks. So much for twenty-six hundred years of pedagogy that assumed that syllogisms were crucial for effective thinking.

The most useful thing I've gained from instruction in categorical reasoning is learning how to draw *Venn diagrams*, named for the nineteenth-century logician John Venn, who invented a pictorial way of representing category membership. Every now and then I find these useful, even necessary, for representing relationships between categories.

Figure 5 shows some of the more useful ones and will give you the general idea.

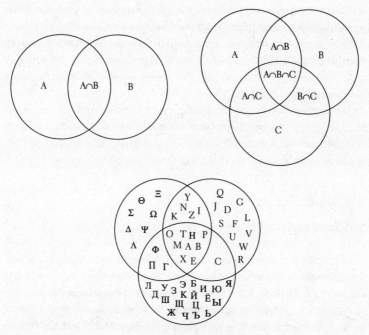

Figure 5. The intersection of categories that overlap with one another.

The top left picture in Figure 5 captures a particular syllogism that we do use in everyday life. It represents a situation in which some (but not all) A are B and some (but not all) B are A. A might stand for small furry animals and B might stand for duck-billed. As it happens there is one animal in the *intersection* of A and B, namely a duck-billed platypus. Or that top left picture could represent a situation in which some but not all of the students who are English speakers at an international school also speak French and some but not all French speakers speak English. (Some but not all A are B and some but not all B are A.) The exclusive English speakers (A only) must study mathematics with Ms. Smith; the exclusive French speakers (B) must study with M. Pirot. Students who speak both languages can study with either teacher.

The picture on the top right shows the much more complicated but not rare situation in which some A are B, some B are A, some A are C, some C are A, some B are C, and some C are B.

The bottom picture presents an actual example of that situation. It represents the intersections of letters that are found in Greek (top left), Latin (top right), and Russian (bottom). I defy you to reach the correct conclusion about overlap of categories solely by means of verbal propositions about categories. In any case, I'm sure I would end up merely with alphabet soup.

This is not enough on Venn diagrams to set you up for a very wide range of problems, but it gives you some of the basics on how to represent category inclusions and exclusions pictorially. You might find that learning more about Venn diagrams will be useful for you.

Propositional Logic

Syllogisms apply only to a tiny range of inferences that we need to make in everyday life. Much more important is propositional logic, which applies to an enormous range of inferential problems. Philosophers and logicians made sporadic contributions to propositional logic between roughly 300 B.C. and A.D. 1300. Beginning in the mid-nineteenth century, logicians started to make considerable progress on propositional logic, in particular focusing on "operators" such as "and" and "or." "And" is concerned with "conjunctions," such as "A is the case and B is the case; therefore A and B are both the case." "Or" is concerned with "disjunctions," such as "A is the case or B is the case; A is the case, therefore B is not the case." Work on propositional logic in that era became the basis for computer design and programming.

At the beginning of the chapter, I asked you to solve a problem about cards. You can now see it was a problem requiring the application of conditional logic. If P then Q. "If a card has a vowel on one side, then it has an even number on the other side." Before we see how well you did on the problem, let's see how you do on the problem below.

You're a police inspector. One of your tasks is to make sure that restaurants don't sell alcohol to anyone under twenty-one. Your task is to say which of the customers below you have to check in order to see whether the customer is obeying this rule: "If a customer is drinking alcohol, the customer is at least twenty-one." You should check only those customers necessary to establish that the rule is being followed.

The first table you see has four customers. You see that

Customer 1	Customer 2	Customer 3	Customer 4
Looks to be over 50	Isn't drinking anything	Is drinking beer	Looks to be under 21

You need to check:

a. Customer 1
b. Customers 1, 2, 3, and 4
c. Customers 3 and 4
d. Customers 1, 3, and 4
e. Customers 1 and 3

I'm betting that you said option c, Customers 3 and 4. Now look back at the card problem. I'm betting that you *didn't* say option c, cards 3 and 4. Can we agree that you should have? The logical structure of the two problems is identical. Check my logic below.

CARD PROBLEM

Make sure this rule is not violated: Vowel? Better be an even number on the other side.

N—doesn't matter whether there's an even number on the back or not.

4—doesn't matter whether there's a vowel on the other side or not.

A—better be an even number on the other side. If not, the rule is broken.

3—better not be a vowel on the other side. If so, the rule is broken.

RESTAURANT PROBLEM

Make sure this rule is not violated: Drinking? Better be twenty-one.

Customer is fifty-plus—doesn't matter whether customer is drinking or not.

Isn't drinking—doesn't matter whether customer is twenty-one or not.

Drinking—better be twenty-one. If not, the rule is broken.

Under twenty-one—better not be drinking. If so, the rule is broken.

Don't feel bad if you didn't get the card problem right. Less than 20 percent of Oxford University students can solve the abstract version of the card problem!

Why is the card problem so much harder than the restaurant problem? At first blush this seems strange, because both problems can be solved by applying conditional logic, in fact by applying the very simplest principle of conditional logic, *modus ponens*:

If P is the case, then Q is the case.	If the customer is drinking, then the customer is twenty-one.
P is in fact the case.	The customer is drinking.
Therefore, Q is the case.	Therefore, the customer is twenty-one.

Modus ponens entails *modus tollens* (if not Q, then not P). An instance where Q (twenty-one or over) is not the case but P (drinking) is the case contradicts the conditional rule.

Note that P (drinking) is a *sufficient condition* but not a *necessary condition* for Q. It's sufficient that P is the case in order for Q to be the case. Lots of other conditions might be sufficient to require that the person is twenty-one. Flying a plane, for example, or gambling.

For the *biconditional*, P is both necessary and sufficient in order for Q to be the case. This would include the (rather weird) rule that if you're drinking you must be twenty-one and if you're twenty-one you must be drinking.

After some more consideration of conditional reasoning, we'll discuss why the drinking problem is so easy.

Plausibility, Validity, and the Logic of the Conditional

As we've seen, syllogistic arguments can be valid—that is, map correctly onto a cogent argument form—even when their conclusions aren't true. That's true as well for propositional logic.

Decide whether each of the following arguments, having two premises and one conclusion, is valid.

ARGUMENT A
Premise 1: If he died of cancer, he had a malignant tumor.
Premise 2: He had a malignant tumor.

_____

Conclusion: Therefore, he died of cancer.

ARGUMENT B
Premise 1: If he died of cancer, he had a malignant tumor.
Premise 2: He didn't die of cancer.

_____

Conclusion: Therefore, he didn't have a malignant tumor.

ARGUMENT C
Premise 1: If he died of cancer, he had a malignant tumor.
Premise 2: He died of cancer.

_____

Conclusion: Therefore, he had a malignant tumor.

Only Argument C is valid. It maps onto modus ponens: If P (he died of cancer) then Q (tumor). P (cancer). Therefore Q (tumor). The plausibility of the conclusions in Arguments A and B tempts us into feeling that they're valid. But Argument A maps onto an invalid argument form: If P (died of cancer) then Q (tumor). Q (tumor). Therefore P (died of cancer). This is called the *converse error* because the form of reasoning involves erroneously converting the premise If P then Q into If Q then P. (If he had a malignant tumor, then he died of cancer.) If that had been the premise, then we would indeed know that since Q is the case, P is also the case. But that wasn't the premise.

We make converse errors all the time—if we're not monitoring our-selves for the logical validity of our argument.

CONVERSE ERROR 1
If the car is not in our garage, then Jane went downtown.
Jennifer told me she saw Jane downtown.
Therefore, the car won't be in the garage.

But of course Jane could have gotten downtown by some means other than taking the car, in which case the car probably *will* be in the garage. Making the error is more likely given some kinds of background knowledge. If Jane rarely goes anywhere without the car, we're more likely to make the error; if she sometimes takes the bus and sometimes is driven by a friend, we're less likely to make the error.

CONVERSE ERROR 2
If I have the flu, then I have a sore throat.
I have a sore throat.
Therefore, I have the flu.

But of course there are possibilities other than P (flu). A cold or strep throat, for example. We're likely to make the error if people are dropping like flies from the flu, always with a sore throat as a symptom, and nothing much else is going around. We're much less likely to make the error if the flu, colds, and pollen allergies are all going around at once.

Argument B above was: If died of cancer, then malignant tumor; didn't die of cancer, therefore didn't have malignant tumor. This is called the *inverse error*. The form of this invalid argument is If P then Q, not P therefore not Q. We make this error a lot, too.

INVERSE ERROR 1
If it's raining, then the streets must be wet.
It's not raining.
Therefore, the streets must not be wet.

If we live in a city where the street sweepers operate frequently (thereby making the streets wet), or if it's a blazing summer day in a city where fire hydrants are sometimes opened to cool people off, we're less likely to make the error. If we live in rural Arizona, with no street sweepers and no fire hydrants, we're more likely to make the error.

INVERSE ERROR 2
If President Obama is Muslim, then he's not a Christian.
President Obama is not Muslim.
Therefore, President Obama is a Christian.

The conclusion would be valid if we had tacit knowledge operating as an additional premise to the effect that people can only be either Muslim or Christian. We don't believe that, of course, but we may have gotten into a mood of thinking these are the only alternatives for Obama; for example, if the only alternatives for Obama's religion ever discussed were Muslim and Christian.

An interesting and important thing to know about the converse and inverse errors is that they are only *deductively* invalid conclusions. (That is, they don't follow logically from their premises.) They may however be pretty good *inductive* conclusions. (That is, if the premises are true, the conclusion is more likely to be true.) It's more likely that I have the flu if I have a sore throat than it is if I don't have a sore throat. If it's not raining, it's less likely that the streets are wet than if it is raining. The plausibility of the inductive conclusion in these cases contributes to making the invalid deductive conclusion plausible as well.

The lists of argument forms and logical errors are very long. But these are some of the most common and important errors.

Pragmatic Reasoning Schemas

The abstract version of the conditional—if P then Q—is difficult to use. We do reason in accord with conditional logic all the time, but rarely by applying a completely abstract version of it. Instead, we're more likely to use what I call *pragmatic reasoning schemas*, that is, sets of rules useful for thinking about everyday life situations.[1] This book is full of such schemas. In fact, at some level, that's what the book is all about. Some of the schemas map directly onto conditional logic. These include, for example, the schema distinguishing between independent and dependent events and the principle that correlation doesn't prove causation. The sunk cost principle and the opportunity cost principle are deductively valid and can be derived logically from the principles of cost-benefit analysis. Economics courses teach these principles, though not as well as they could because they're typically not very good at showing how the formal principles can be used pragmatically for everyday reasoning.

Some pragmatic reasoning schemas map onto conditional logic but fall short of being deductively valid because they don't guarantee a correct answer. In fact, they're not concerned with truth or validity at all but with assessing whether a person's conduct is proper. This branch of logic is called *deontic*, from the Greek *deon*, meaning duty. It deals with what kind of situation constitutes an obligation, what kind gives permission, what's optional behavior, what's beyond the call of duty, and what ought to be done. *Contractual schemas* are one type of deontic schema that can be used to solve a wide range of problems related to permission and obligation.

The deontic schema that's necessary for getting the drinking age problem right is called the *permission schema*.[2] You want to drink (P)? Better be twenty-one (Q). Not twenty-one (not Q)? Better not be drinking (not P).

A kindred schema is the *obligation schema*.[3] If you're eighteen (P), you must register for the draft (Q). Didn't register for the draft (not Q)? Then you better not be eighteen or you didn't meet your obligation.

Two years of law school improves deontic reasoning quite a bit, but two years of graduate training in philosophy, psychology, chemistry, or medicine does nothing for this kind of reasoning.[4]

A second type of pragmatic reasoning schema doesn't map onto conditional logic at all (or at least it isn't very profitable to try such a mapping), but applies to an enormous range of situations and can be described in purely abstract terms. Logical thinking is required for using these schemas, but the logic isn't what makes them powerful; rather, it's their ability to shed light on everyday problems. These include statistical schemas and schemas for scientific procedures such as the randomized control design. Statistics and methodology courses teach these concepts but don't always succeed very well in creating pragmatic schemas that are helpful in everyday life. Both undergraduate and graduate courses in social sciences and psychology, but not natural sciences or the humanities, do enhance pragmatic schemas that help in applying statistical and methodological schemas to everyday problems.[5] Other highly general pragmatic reasoning schemas include Occam's razor, the tragedy of the commons, and the concept of emergence, discussed in Chapter 15.

Finally, some powerful pragmatic reasoning schemas don't constitute abstract blueprints for reasoning but are merely empirical principles that facilitate correct solutions to a broad range of everyday problems. These include the fundamental attribution error, the generalization that actors and observers tend to explain behavior differently, loss aversion, the status quo bias, the principle that some choice architectures are generally superior to others in the quality of choices they encourage, and the principle that incentives aren't necessarily the best way to get people to change their behavior—among dozens of others in this book.

Abstract pragmatic schemas are tremendously useful, but purely logical schemas are of limited value. I believe this is the case because there's a very high civilization, namely Confucian Chinese, that never developed purely logical formalisms. It's that civilization's dialectic tradition, and modern additions to it, that are covered in the next chapter.

Summing Up

Logic divests arguments of any references to the real world so that the formal structure of an argument can be laid bare without any interference from prior beliefs. Formal logic, contrary to the opinions of educators for twenty-six hundred years, doesn't constitute the basis of everyday thought. It's primarily a way of thinking that can catch some kinds of errors in reasoning.

The truth of a conclusion and the validity of a conclusion are entirely separate things. The conclusion of an argument is valid only if it follows logically from its premises, though it may be true regardless of the truth of the premises or whether it follows logically from the premises. An inference need not be logically derivable from any other premises, but it gains in claims for credence if it can be shown to have logical as well as empirical support.

Venn diagrams embody syllogistic reasoning and can be helpful or even necessary for solving some categorization problems.

Errors in deductive reasoning are sometimes made because they map onto argument forms that are inductively valid. That's part of the reason we're susceptible to making deduction errors.

Pragmatic reasoning schemas are abstract rules of reasoning that underlie much of thought. These include deontic rules such as the permission schema and the obligation schema. They also include many inductive schemas discussed in this book such as those for statistics, cost-benefit analysis, and reasoning in accord with sound methodological procedures. Pragmatic reasoning schemas are not as general as the rules of logic because they apply only in specific situations, but some of them rest on logical foundations. Others, such as Occam's razor and the concept of emergence, are widely applicable but don't rest on formal logic. Still others are merely empirical generalizations of great practical utility, such as the fundamental attribution error.

14. Dialectical Reasoning

The most striking difference between the traditions at the two ends of the civilized world is in the destiny of logic. For the West, logic has been central and the thread of transmission has never snapped.

—Angus Graham, philosopher

It is precisely because the Chinese mind is so rational that it refuses to become rationalistic and . . . separate form from content.

—Shu-hsien Liu, philosopher

To argue with logical consistency . . . may not only be resented but also be regarded as immature. —Nobuhiro Nagashima, anthropologist

If you're a person who grew up in a Western culture, you may be surprised by the fact that one of the world's great civilizations, namely China, has no history of formal logic.

From before the time of Plato until very recent times, when the Chinese encountered Western thought, there was virtually no concern at all with logic in the East.[1] At the moment when Aristotle was developing formal logic, the Chinese philosopher Mo-tzu and his followers did deal with some issues touching on logic, but neither he nor anyone else in the tradition of classical Chinese culture ever developed a formalized system.[2] And after a brief flurry of interest in Mo-tzu's thought, the trail of logic went cold in the East. (Mo-tzu, incidentally, did systematic work on cost-benefit analysis, centuries before anyone in the West dealt seriously with the topic.)[3]

So how did the Chinese manage to make great progress in mathematics, as well as invent hundreds of important things that the West invented much later or not at all, if they lacked a tradition of logic?

We're forced to acknowledge that a civilization can make enormous strides without ever paying much attention to formal logic. This is true not only of China but of all cultures in East Asia with roots in the Confucian tradition, including Japan and Korea. It's not true of India, where there is a concern with logic extending back to roughly the fifth or fourth century B.C. Interestingly, the Chinese were aware of Indian work on logic and translated some logic texts by Indians. But the Chinese translations are full of errors, and the influence of these texts was minimal.

The system of thought that the Chinese developed instead of logic has been called *dialectical reasoning*. Dialectical reasoning is actually opposed to formal logic in many ways.

Western Logic Versus Eastern Dialecticism

Aristotle placed at the foundation of logical thought the following three propositions.

1. Identity: A = A: Whatever is, is. A is itself and not some other thing.
2. Noncontradiction: A and not A can't both be the case. Nothing can both be and not be. A proposition and its opposite can't both be true.
3. Excluded middle: Everything must either be or not be. A or not A can be true but not something in between.

Modern Westerners accept these propositions. But people who grew up in the intellectual tradition of China don't buy into them—at least not for every kind of problem. Instead, it is dialecticism that is at the foundation of Eastern thought.

As the psychologist Kaiping Peng has described, three principles underlie Eastern dialecticism.[4] Notice I didn't say "propositions." Peng warns that the term "proposition" has much too formal a ring for what is a generalized stance toward the world rather than a set of ironclad rules.

1. *Principle of change*:
   Reality is a process of change.
   What is currently true will shortly be false.
2. *Principle of contradiction*:
   Contradiction is the dynamic underlying change.
   Because change is constant, contradiction is constant.
3. *Principle of relationships (or holism)*:
   The whole is more than the sum of its parts.
   Parts are meaningful only in relation to the whole.

These principles are intimately linked. Change creates contradiction and contradiction produces change. Constant change and contradiction imply that it's meaningless to discuss the individual part without considering its relationships with other parts and with prior states of the world.

The principles also imply another important tenet of Eastern thought, which is the insistence on finding the "middle way" between extreme propositions. There is a strong presumption that contradictions often are merely apparent, and an inclination to believe that "A is right but not A is not wrong." This stance is captured by the Zen Buddhist dictum that "the opposite of a great truth is also true."

To many Westerners, these notions may seem reasonable and even familiar. The Socratic dialogue, often called dialectical, is similar in some ways. This is a conversation exchanging different viewpoints, with the goal of more closely approaching the truth. Jews borrowed that version of dialectical thinking from the Greeks, and Talmudic scholars developed it over the next two millennia and more. Western philosophers of the eighteenth and nineteenth centuries such as Hegel and Marx made contributions to the dialectical tradition. Dialectical reasoning became a topic of serious study by cognitive psychologists—in both the East and the West—beginning in the late twentieth century.

The Eastern dialectical stance reflects the deep infusion of Eastern thought by the Tao. The Tao means a thousand things to an Easterner, but at base it captures the concept of change. Yin (the feminine and dark and passive) alternates with yang (the masculine and light and active). Indeed, yin and yang only exist because of each other, and when

Figure 6. The sign of the Tao.

the world is in a yin state, this is a sure sign that it's about to be in a yang state. The sign of the Tao, which means the "Way" to exist with nature and with one's fellow humans, consists of two forces in the form of a white and a black swirl.

The concept of change is represented by the fact that the black swirl contains a white dot and the white swirl contains a black dot. And "the truest yang is the yang that is in the yin." The yin-yang principle expresses the relationship that exists between opposing but interpenetrating forces that may complete each other, make each other comprehensible, or create the conditions for one to alter into the other.

From the *I Ching* (*Book of Changes*): "For misery, happiness is leaning against it; for happiness, misery is hiding in it. Who knows whether it is misery or happiness? There is no certainty. The righteous suddenly becomes the vicious, the good suddenly becomes the bad."

A familiarity with Eastern dialecticism makes it easier to understand the very different assumptions about change characteristic of Eastern and Western thought. Li-Jun Ji has shown that for trends of any kind at all—world tuberculosis rate, growth in GDP of developing countries, rates of diagnosis of autism in American children—Westerners tend to assume the trend will continue in its current direction; Easterners are much more likely to assume that a given trend may level off or actually reverse itself.[5] Business school students steeped in the Western tradition are inclined to buy a stock that's going up and dump a stock that's going down.[6] Students raised in the Eastern tradition are inclined to buy a stock that's going down and sell one that's going up. (Recall from Part II that this is a clear instance of an erroneous preference.)

The dialectical tradition explains in part why East Asians are more attentive to context (discussed in Chapter 2). If things are constantly changing, you better pay attention to the circumstances surrounding a given event. Things are going on to influence the event that will result in change and contradiction.

The logical tradition and the dialectical tradition produce quite different reactions to contradictory propositions and arguments. If you present people with two propositions implying the opposite of each other—which are close to being a direct contradiction—Westerners and Easterners respond in very different ways. Students at the University of Michigan and at Beijing University were presented with pairs of alleged scientific findings.[7] For example, some students read: (1) Fuel usage in a large number of developing countries indicates a great worsening of environmental problems including global warming, and (2) A meteorologist studied temperatures in twenty-four widely separated parts of the world and found that temperatures had actually dropped by a fraction of a degree each of the last five years. Other students read just one of these propositions. All students were asked how plausible they found the propositions to be.

Michigan students were more inclined to believe the more plausible proposition, such as (1) above, when they saw it contradicted by a less plausible proposition, such as (2) above, than when they merely saw the more plausible proposition by itself. This pattern is not logically *coherent*. A proposition can't be more believable when it's contradicted by something than when it's not contradicted. The error probably occurs because of Western eagerness to resolve a contradiction by deciding which proposition is correct. The process of choosing involves focusing on all the reasons the more plausible proposition is to be preferred. Confirmation bias at work. Thus beefed up, the more plausible proposition seems stronger than it would if the person hadn't gone through the process of choosing between it and a seemingly contradictory, less plausible proposition.

Chinese students' behavior couldn't be more different. They placed more credence in the *less* plausible proposition when they saw it contradicted than when they didn't see it contradicted. This is also logically incoherent, but follows from the sense that there must be some truth to each of two contradictory statements. Once the less probable proposition has been bolstered by finding ways in which it might be true, it seems more plausible than if that bolstering process hadn't occurred. One might almost say that Easterners sometimes display an anticonfirmation bias!

So Western thought can get things wrong in its rush to stamp out a seeming contradiction rather than entertaining the possibility that both propositions might have some truth. Eastern thought can get things wrong by finding weak propositions more plausible when contradicted because of an attempt to bolster a weak proposition in order to split the difference with a contradictory but stronger argument.

The logical and dialectical systems of thought have a lot to learn from each other: each gets some things right that the other gets wrong.

Logic Versus the Tao

The shaky hold on logic characteristic of East Asians is evident today in the thinking even of young people studying in the best Asian universities.

Consider the three arguments below. Which seem to you to be logically valid?

ARGUMENT 1
Premise 1: No police dogs are old.
Premise 2: Some highly trained dogs are old.

Conclusion: Some highly trained dogs are police dogs.

ARGUMENT 2
Premise 1: All things that are made of plants are good for the health.
Premise 2: Cigarettes are things that are made of plants.

Conclusion: Cigarettes are good for the health.

ARGUMENT 3
Premise 1: No A are B.
Premise 2: Some C are B.

Conclusion: Some C are not A.

The first argument is meaningful and has a plausible conclusion, the second argument is meaningful but its conclusion is not plausible, and the third argument is so abstract that it makes no contact at all with any real-world facts. Despite the plausibility of its conclusion, argument 1 is invalid. Despite the implausibility of argument 2, it is valid. And the meaningless argument 3, as it happens, is valid. (Try drawing Venn diagrams of these arguments to see how helpful they can be in assessing validity.)

The psychologists Ara Norenzayan, Beom Jun Kim, and their co-workers attempted to find out whether Asians and Westerners thought about problems like those above differently. They presented Korean and American university students with arguments that were either valid or invalid and that had conclusions that were either plausible or implausible.[8] The researchers asked them to evaluate whether the conclusion followed logically from the premises for each argument. There were a total of four different types of syllogisms, ranging from very simple structures to quite complicated ones.

Both Koreans and Americans were more likely to rate syllogisms with plausible conclusions as being valid, whether they were or not. But Koreans were much more influenced by the plausibility of the conclusion than were Americans. This didn't happen because the Korean participants were less capable of performing logical operations than the American participants. The two groups made an equal number of errors on the purely abstract syllogisms. It's just that Americans are more in the habit of applying logical rules to ordinary events than are Koreans and are therefore more capable of ignoring the plausibility of the conclusion.

East Asian university students also make mistakes in syllogisms based on how typical the member of a category is. The investigators told the students, for example, that all birds have a given property (a made-up one, such as "have an omentum"). They next asked the students either how convincing it is that eagles have the property, or how convincing it is that penguins have the property. The two conclusions are of course equally valid. Americans were much less affected by typicality than were Koreans. For example, Koreans were not as convinced as Americans were that penguins have the property given that birds have the property.

Finally, East Asian students have more problems with propositional logic than do Americans. They're more thrown off by their desires. If they would like a particular conclusion to be true, they're more likely to judge incorrectly that the conclusion follows from the premises.[9] This is not the sort of error one wants to make. It indicates that facility with logic—being able to strip meaning away from propositions and convert them into abstractions—helps Westerners avoid undue influences on judgments.

Context, Contradiction, and Causality

Recall from Chapter 2, in the discussion about the importance of context, that the Western approach to thinking about the world is to focus on a central object (or person). Westerners identify the attributes of an object, assign the object to a category, and apply the rules that govern that category of objects. The underlying purpose is often to establish a causal model of the object so that it can be manipulated for one's own goals.

The Eastern approach is to attend much more broadly to the object in its context, to the relationships among objects, and to the relationship between object and context.

Different approaches to historical analysis follow from these differences in how best to understand the world. Japanese teachers of history begin by setting out the context of events in some detail.[10] They then proceed through the important events in chronological order, linking each event to its successor. Teachers encourage their students to imagine the mental and emotional states of historical figures by thinking about the analogy between their situations and the situations of the students' everyday lives. The actions are then explained in terms of these feelings. Teachers regard students as having good ability to think historically when they show empathy with the historical figures, including those who were Japan's enemies. "How" questions are asked frequently—about twice as often as in American classrooms.

American teachers spend less time setting the context than Japanese teachers do. They begin with the *outcome*, rather than with the initial event or catalyst. The chronological order of events is slighted or even destroyed in presentation. Instead, the order in which factors are considered is dictated by discussion of the causal factors assumed to be important. ("The Ottoman empire collapsed for three major reasons.") Teachers regard students as having good ability to reason historically when they are capable of providing evidence to fit their causal model of the outcome. "Why" questions are asked twice as frequently in American classrooms as in Japanese classrooms.

Both approaches seem useful—and complementary. But in fact, East Asian historical analysis seems simply wrong to Westerners. And in general, rather than appreciating the holistic style of thought characteristic of the East, Westerners frequently reject the approach. Remarkably, the children of Japanese businessmen living in the United States are sometimes put back a grade in their American school because their teachers find them to be lacking in analytic ability.

The different types of thought result in very different *metaphysics*, or assumptions about the nature of the world. The differences in patterns of thought also produce different physics. Because of their attention to context, the ancient Chinese got some things right that the Greeks got wrong.

Ancient Chinese attention to context encouraged the realization that there could be action at a distance. This allowed the Chinese to correctly understand questions of acoustics and magnetism, as well as to comprehend the true reason for the tides. That the moon could drag the oceans along eluded even Galileo.

Aristotle's explanation of why objects fall when dropped into water is that they possess the property of gravity. But not all objects fall when dropped into water; some stay on the surface. Those objects, Aristotle explained, possess the property of levity. But of course there is no such property as levity, and gravity is a relation between objects rather than a property of a single object.

Einstein had to add a cheater factor to his theories about the nature of the universe, namely, the cosmological constant, in order to account for what he was confident was the steady state of the universe. But of course the universe isn't in a steady state as had been assumed since the time of Aristotle. As a Westerner thoroughly imbued with Greek assumptions about stasis, Einstein had a gut instinct that the universe should be constant, so he resorted to the cosmological constant to preserve the assumption.

Chinese dialectical reasoning had an impact on the physicist Niels Bohr, who was highly knowledgeable about Eastern thought. He attributed his development of quantum theory in part to the metaphysics of the East. There had been a centuries-long debate in the West about whether light consists of particles or waves. Belief in one was assumed to contradict and render impossible belief in the other. Bohr's solution was to say that light can be thought of in both ways. In quantum theory, light can be viewed either as a particle or as a wave. Just never both at the same time.

But while the Chinese got a lot of things right that the West got wrong, they could never have proved their theories correct. For that it takes science, which the West has had for twenty-six hundred years. Science, at base, is categorization plus empirical rules and a commitment to logical principles. The Chinese understood the concept of action at a distance when Westerners didn't, but it was Western science that proved the concept to be correct. This was done by scientists who set out to conduct experiments intended to show that action at a distance was impossible! They were startled by findings showing it was indeed possible.

Stability and Change

There is a profound difference between Eastern and Western beliefs about change. For reasons that aren't at all clear to me, the Greeks were positively obsessed with the idea that the universe and the objects in it don't change.

It's true that Heraclitus and other sixth-century B.C. philosophers recognized that the world changes. ("A man never steps in the same river twice because the man is different and the river is different.") But by the fifth century B.C., change was out and stability was in. Heraclitus's views were actually ridiculed. Parmenides "proved," in a few easy steps, that change was impossible: To say of a thing that it doesn't exist is a contradiction. Nonbeing is self-contradictory and so nonbeing can't exist. If nonbeing can't exist, then nothing can change, because if thing 1 were to change to thing 2, then thing 1 would not be!

Zeno, Parmenides's pupil, proved to many Greeks' satisfaction that motion is impossible. One proof is his famous treatment of the arrow.

1. When an arrow is in a place just its own size, it's at rest.
2. At every moment of its flight, the arrow is in a place just its own size.

3. Therefore, at every moment of its flight, the arrow is at rest.
4. Since the arrow is always at rest, we see that motion (hence change) is impossible.

Another of Zeno's proofs is the Achilles Paradox. If Achilles is trying to catch a slower runner ahead of him—a tortoise, say—he must run to the place where the tortoise is at the moment. But by the time Achilles gets *there*, the tortoise has moved on. Ergo, Achilles can never catch the tortoise. Since fast runners can never catch slow ones, we can deduce that motion never occurs.

As the communications theorist Robert Logan has written, the Greeks were enslaved to the rigid linearity of their either/or logic.[11]

The Greek insistence on an unchanging or highly stable world echoes down through the centuries. The extreme Western insistence on attributing human behavior to a person's enduring dispositions rather than to situational factors—the fundamental attribution error—is directly traceable to Greek metaphysics.

One of the clearest examples of the damage done by the fundamental attribution error has to do with Western (mis)understanding of some important influences on intelligence and academic achievement.

I began to have trouble with math in the fifth grade. My parents assured me that was to be expected: Nisbetts had never been much good at math. I was delighted to have the alibi. But in retrospect I can see that my problems with math began after a two-week bout of mononucleosis sidelined me from school. As it happened, that was the period when my class had taken up fractions. I'm still not much good at math, though I do believe I would have been better at it if I hadn't accepted my parents' dispositional inference about my ability to deal with fractions.

Contrast my parents' attribution with what might be expected from a Chinese American tiger mother. "You come home with Bs in math on your report card? You start getting As if you want to be a member of this family!"

For the two thousand years since it became possible for a Chinese peasant boy to become the most powerful magistrate in the land through study, the Chinese have believed that hard work makes you smart. Confucius believed that part of ability was "a gift from Heaven," but most of it was due to effort.

A study of American high school seniors begun in 1968 found that the Chinese-heritage students among them were about even in IQ with their Caucasian counterparts.[12] But the students of Chinese heritage scored about a third of a standard deviation better on the SAT. SAT scores are highly correlated with IQ, but SAT scores owe more to study than IQ scores do. Astonishingly, a couple of decades after high school graduation, the Chinese Americans were 62 percent more likely to work in professional, managerial, or technical fields than the European Americans.[13] Even among European Americans, students who view ability as modifiable do much better in school than those who don't.[14] And when European Americans are taught that how smart you are is in good part due to how hard you work, it improves their school performance. Learning about the importance of effort is particularly powerful for poor black and Hispanic children.[15]

East-West differences in beliefs about malleability and change reverberate through an enormous number of domains of life. People of European culture—especially Americans—label a person convicted of theft or homicide as a "criminal." Asians avoid such firm categorizations. Perhaps as a consequence, lengthy incarcerations are relatively rare in Asia. The U.S. incarceration rate is five times that of Hong Kong, eight times that of South Korea, and fourteen times that of Japan.

Dialecticism and Wisdom

The letter below was sent to the advice columnist Abigail Van Buren and published in scores of newspapers. Please think for a moment about the likely outcome of a situation like the one described there.

Dear Abby:

My husband, "Ralph," has one sister, "Dawn," and one brother, "Curt." Their parents died six years ago, within months of each other. Ever since, Dawn has once a year mentioned buying a headstone for their parents. I'm all for it, but Dawn is determined to spend a bundle on it, and she expects her brothers to help foot the bill. She recently told me she had put $2,000 aside to pay for it. Recently Dawn called to announce that she had gone ahead, selected the design, written the epitaph and ordered the headstone. Now she expects Curt and Ralph to pay "their share" back to her. She said she went ahead and ordered it on her own because she has been feeling guilty all these years that her parents didn't have one. I feel that since Dawn did this all by herself, her brothers shouldn't have to pay her anything. I know that if Curt and Ralph don't pay her back, they'll never hear the end of it, and neither will I. What should I do about this?

After a little more explication of East-West differences in thinking patterns, we'll return to this vignette.

Recall that Jean Piaget, the great mid-twentieth century developmental psychologist, held that the basis of all thought after childhood was propositional logic. He referred to these logical rules as "formal operations" as distinct from the "concrete operations" that characterize how children think about specific real events, such as the conservation of matter across the shape of its container. (There's neither more nor less sand when you pour it from a tall narrow container into a short wide one.) Piaget held that young children use logic to develop their understanding of events in the world, but they lack the ability to use logic to think abstractly. As children enter adolescence they shift toward using formal operations to think about abstract concepts. Formal operations—the rules of propositional logic—can only be induced and can't be taught. They are fully developed by the end of adolescence. There is no more learning about how to think using abstract rules after that point. Every normal adult has the exact same set of rules for formal logic.

Most of that story is mistaken. As this book shows, there are innumerable abstract rules beyond those of formal operations, such as the concepts of statistical regression and cost-benefit analysis. Moreover, these abstract rules can be both induced and taught, and we keep on learning them well after adolescence. Partly as a reaction to Piaget's theories, psychologists in the late twentieth century began to define what they called "postformal operations," that is to say, principles of thought that are learned primarily after adolescence and typically don't guarantee a single correct answer but rather a range of plausible answers. Instead, application of the principles may result merely in new perspectives on problems or provide pragmatic guidance for dealing with apparent logical contradictions and social conflict.

The postformalists, notably Klaus Riegel and Michael Basseches, labeled this type of thinking "dialectical."[16] They relied heavily on Eastern thought in describing and elucidating these principles, which can be grouped under five rubrics.

Relations and context. Dialectical thinking emphasizes attention to relations and contexts, the importance of locating an object or phenomenon as part of some larger whole, an emphasis on understanding how systems function, a concern with equilibrium in systems (such as the body, groups, factory operations), and the need to view problems from many perspectives.

Antiformalism. Dialectical thinking opposes formalism because of its separation of form from content. We make errors by abstracting the elements of a problem into a formal model and ignoring facts and contexts crucial to correct analysis. Overemphasis on logical approaches leads to distortion, error, and rigidity.

Contradiction. The postformalists emphasize the importance of identifying contradictions between propositions and between systems and of recognizing that opposites can complement each other and lead to greater understanding than an insistence on the rejection of one idea in favor of another.

Change. Postformalist psychologists emphasize the importance of understanding events as moments in a process rather than static, one-off occurrences. They recognize interaction between systems as a source of change.

Uncertainty. Partly because of their emphasis on change, their acknowledgment of contradiction, and their recognition of the multiple influences present in most contexts, postformalists place a value on recognizing the uncertainty of knowledge.

These principles of thought are not alien to Westerners. The difference between Easterners and Westerners is that Easterners regard them as fundamental and use them all constantly. Let's look at some examples of the use of these principles for everyday problems.

Culture, Aging, and Dialecticism

The psychologists Igor Grossmann, Mayumi Karasawa, Satoko Izumi, Jinkyung Na, Denise Park, Michael Varnum, Shinobu Kitayama, and I gave problems such as the Dear Abby dilemma described a few pages back, as well as problems concerning societal conflicts such as ethnic discord and disagreements about use of natural resources, to people of a wide variety of ages and social classes in both the United States and Japan.[17] We asked participants what they thought would happen next and why, and coded answers in terms of six categories related to dialectical reasoning.

1. Does the answer avoid rigid application of a rule?
2. Does the answer take into account the perspectives of each protagonist?
3. Is the answer attentive to the nature of the contradictory views?
4. Does the answer recognize the possibility of change rather than a stalemate?
5. Does the answer mention possible forms of compromise?
6. Does the answer express uncertainty rather than dogmatic confidence?

We found that young and middle-aged Japanese responded to interpersonal and societal conflicts in a more dialectical fashion than did young and middle-aged Americans.[18] The Japanese were more likely to avoid rigid application of a rule, more likely to take into account the perspectives of all participants, more attentive to the nature of the conflict, and more likely to recognize the possibility of change and compromise. They expressed less certainty about their conclusions.

Table 5 gives examples of answers reflecting more and less dialectical approaches to the Dear Abby column describing a conflict among siblings over a payment for a headstone for their mother. All the answers are those of American subjects, but Japanese answers were completely comparable—it's just that they were more likely to provide dialectical answers.

TABLE 5. EXAMPLES OF RESPONSES FOR THE HEADSTONE STORY SHOWING MORE AND LESS DIALECTICAL REASONING

Less Dialectical	*More Dialectical*
Considering the Different Perspectives of People Involved in the Conflict	
I can imagine that it was a sour relationship afterward because let's just say that Kurt and Ralph decided not to go ahead and pay for the headstone. Then it's going to create a gap of communication between her sister and her brothers. If the gravestone was just as important to them, then it wouldn't have been a problem about them getting the money in the beginning.	Somebody might believe that we need to honor parents like this. Another person might think there isn't anything that needs to be done. Or another person might not have the financial means to do anything. Or it could also mean that it might not be important to the brothers. It often happens that people have different perspectives on situations important to them.
Recognizing Multiple Ways the Conflict Might Unfold	
She probably ended up having to pay for it by herself and she probably bugs them about it. Because if they wanted to help they would have already given her money, I think. I don't think there really is an outcome.	It could have several outcomes. The brothers might have reimbursed the sister and then there was resentment on the wife's part. Or there could have been resentment for all three. Or the brothers could have refused to pay and she may have accepted it. Or maybe one brother would have paid.

(continued)

TABLE 5. CONTINUED

Less Dialectical	*More Dialectical*
Search for a Compromise	
They probably didn't have the money; otherwise they would have done it sooner. And Dawn's going to be stuck with a bill, period. She should be stuck with a bill if she went ahead without their okay. I think she footed the bill herself and so she was bitter toward her brothers after that, which she shouldn't be. She took it upon herself.	I would think there would probably be some compromise reached, that Kurt and Ralph realize that it's important to have some kind of headstone. Although Dawn ordered it without them agreeing, they would probably pitch in somehow, even if it was not what she wanted ideally. But hopefully there was some kind of contribution.

In our view, the generally more dialectical answers of the Japanese were a reflection of greater wisdom on their part. And we're in good company. We presented the problems, along with Japanese and American answers, to members of the University of Chicago–based Wisdom Network. The network consists of (heavily Western) philosophers, social psychologists, psychotherapists, and members of the clergy interested in the nature of wisdom and how people can attain it. Members of the network endorsed the more dialectical answers to the Dear Abby–type problems as being wiser.

As people get older, do they get wiser in the sense that they become more likely to apply dialectical reasoning to social conflict? Americans do. From the age of about twenty-five to about seventy-five, Americans get steadily more likely to apply dialectical approaches to interpersonal and societal problems.[19]

It stands to reason that people would become wiser about how to handle social conflict as they get older. They could be expected to become more likely to recognize the potential for it, learn ways of avoiding it, and develop ways of reducing it if it occurs.

But Japanese don't get wiser in these respects.

Here's how we account for the fact that Americans become more dialectical with age and Japanese don't. Younger Japanese are wiser about conflict than younger Americans because their socialization emphasizes attention to social context. They are explicitly taught how to avoid and reduce conflict, which is much more damaging to the social fabric in the East than in the West.

Young Americans are less likely to be taught dialectical principles or how to deal with conflict. But as Americans experience more and more conflict over their life span, they induce ever-better ways of recognizing and dealing with it. Japanese don't get better with age because they're merely applying early-learned concepts rather than increasing their repertoire of conflict-related principles in the conduct of everyday life. Moreover, they encounter much less conflict in their daily lives than Americans, and so they don't have occasion to induce ever-better ways of dealing with it.

So is logical thinking or dialectical thinking better in general? This may sound like a nonsensical question. I've argued that both have their advantages and disadvantages. Sometimes it's helpful to abstract an argument to the point that its logical structure can be examined, but sometimes it's a mistake to insist on separating form from content. Sometimes it's helpful to try to dissolve contradictions, but sometimes it's more productive to acknowledge them and see whether the truth might lie between the contradictory ideas, or whether it's possible to transcend the contradictions and find some respect in which both are true.

But I'll stick my neck out and hazard the generalization that logical thinking is crucial for scientific thought and some kinds of well-defined problems. Dialectical thinking is often more helpful for thinking about everyday problems, especially those involving human relations.

Assuming that you agree with East Asians, the elderly, and the members of the Chicago Wisdom Network about the value of dialectical thinking, could you learn to be more dialectical in your own life?

I think so. And I think you have already. Much of what you've read in this book is sympathetic to dialectical reasoning and skeptical about too much reliance on formal analytic procedures. The book has emphasized the importance of paying attention to context (thereby combatting the fundamental attribution error), the likelihood of variability and change in processes and in individuals (weakening susceptibility to the interview illusion), the fact that the attributes of objects and people tend to be associated with other attributes (encouraging attention to problems of self-selection), and the uncertainty of knowledge (strengthening concern with true score, measurement error, accuracy of assessment of correlations, reliability, and validity). And most important of all: ATTBW—Assumptions Tend to Be Wrong.

Summing Up

Some of the fundamental principles underlying Western and Eastern thought are different. Western thought is analytic and emphasizes logical concepts of identity and insistence on noncontradiction; Eastern thought is holistic and encourages recognition of change and acceptance of contradiction.

Western thought encourages separation of form from content in order to assess validity of arguments. A consequence is that Westerners are spared some logical errors that Easterners make.

Eastern thought produces more accurate beliefs about some aspects of the world and the causes of human behavior than Western thought. Eastern thought prompts attention to the contextual factors influencing the behavior of objects and humans. It also prompts recognition of the likelihood of change in all kinds of processes and in individuals.

Westerners and Easterners respond in quite different ways to contradictions between two propositions. Westerners sometimes actually believe a strong proposition more when it is contradicted by a weak proposition than when encountering it by itself. Easterners may actually believe a weak proposition more when it is contradicted by a strong proposition than when encountering it by itself.

Eastern and Western approaches to history are very different. Eastern approaches emphasize context, preserve the order of events and emphasize the relations between them, and encourage empathy with historical figures. Western approaches tend to slight contextual factors, are less concerned about preservation of the sequence of events, and emphasize causal modeling of historical processes.

Western thought has been influenced substantially by Eastern thought in recent decades. Traditional Western propositional logic has been supplemented by dialectical principles. The two traditions of thought provide good platforms for critiquing each other. The virtues of logical thought seem more obvious in light of dialectical failings, and the virtues of dialectical thought appear more obvious in light of the limitations of logical thought.

Reasoning about social conflict by younger Japanese is wiser than that of younger Americans. But Americans gain in wisdom over their life span and Japanese do not. Japanese, and undoubtedly other East Asians, are taught about how to avoid and resolve social conflict. Americans are taught less about it and have more to gain as they grow older.

PART VI

KNOWING THE WORLD

Many years ago, I began having casual conversations about reasoning with two young philosophers, Stephen Stich and Alvin Goldman. The conversation turned serious when we began to realize that we were interested in many of the same questions concerning *epistemology*. Epistemology is the study of what counts as knowledge, how we can best obtain knowledge, and what can be known with certainty. The three of us and a psychology graduate student named Tim Wilson started a long-running seminar.

The philosophers were quite taken with the idea that there was a science purporting to address empirically some of the philosophical questions about knowledge that had been around for twenty-six hundred years. They were intrigued to find that psychologists had begun to study reasoning tools of the kind reported in this book, such as schemas and heuristics, and to show the relevance of tools of scientific discovery to understanding of everyday life. Moreover, they saw that psychologists really did have ways of scientifically investigating some of these questions. They also saw that the philosophical literature had much to offer the scientific approach to reasoning, both with respect to guidance about the important questions to ask and what can be regarded as knowledge.

Goldman gave the name "epistemics" to the new discipline that fuses theory of knowledge, cognitive psychology, and philosophy of science (which is concerned with appraisal of the methods and conclusions of scientists). Stich began a movement called X ϕ. The X stands for "experimental"; the Greek symbol *phi* stands for "philosophy." Stich and his many students have continued to do work that is both excellent psychology and important philosophy. I hasten to say that none of us was as original as we initially thought. It turned out that many philosophers and psychologists were thinking along similar lines. But I think we did help to crystallize some important ideas that were floating around in the zeitgeist.

Chapters 15 and 16 deal in part with epistemics as defined by Goldman. The work discussed also reflects the experimental stance of X ϕ as developed by Stich. The philosopher's stock-in-trade has always been assertions about "our intuitions." Stich and his colleagues have shown that intuitions about the nature of the world, what one can call knowledge, and what one regards as moral can be so diverse across cultures and from individual to individual that it often makes no sense to appeal to a chimera called "our intuitions."[1]

15. KISS and Tell

We consider it a good principle to explain the phenomena by the simplest hypothesis possible. —Claudius Ptolemy

It is futile to do with more things that which can be done with fewer. —William of Occam

To the same natural effects we must, so far as possible, assign the same causes. —Isaac Newton

Whenever possible, substitute constructions out of known entities for inferences to unknown entities. —Bertrand Russell

What counts as knowledge, and what qualifies as an explanation, are two of the main questions discussed in this book. They are also central concerns for philosophers of science. In the answers they give to these questions, philosophers of science perform both the function of describing what it is that scientists are doing and critiquing what they do. Conversely, some philosophers of science use the findings of scientists—and experimental philosophers—to address traditional philosophical questions (although this is a more controversial practice among philosophers than you might guess).

Some of the important issues addressed by philosophers of science include: What constitutes a good theory? How economical or simple should a theory be? Can a scientific theory ever be confirmed, or is "not yet falsified" the best it can do? Can a theory be a good one if there is no way we could falsify it? What's wrong with special-purpose, "ad hoc" fixes to theory? All of these questions are as relevant to the theories and beliefs we hold about everyday life events as they are to the activities of scientists.

KISS

In graduate school, I had a professor who was prone to generating highly complicated theories—much more complicated than I thought testable or likely to be supported by evidence in a convincing way. He defended himself by saying, "If the universe is pretzel-shaped, you better have pretzel-shaped hypotheses." My response, prudently uttered only to myself, was, "If you start out with pretzel-shaped hypotheses, the universe better be pretzel-shaped or you'll never find out what shape it is. Better to start with a straight line and go from there."

The injunction against complexity has come to be labeled Occam's razor: theories must be succinct—unnecessary concepts have to be shaved away. In the scientific arena, the simplest theory capable of explaining the evidence wins. We abandon a simple theory only when there's a more complicated theory that explains more evidence than the simple theory. Simpler theories are also to be preferred because they tend to be easier to test, and, in the more precise sciences, more readily modeled mathematically.

Ptolemy didn't follow his own advice very well. Figure 7 shows the path of Mars around the earth as specified by Ptolemy, providing epicycle after epicycle in order to match the perceived motion of Mars. An epicycle is a circle on a circle. There was a strong prior assumption in Ptolemy's day that the universe was constructed on elegant geometric principles, employing especially the circle. If lots of circles were required to model planetary movement, so be it.

Ptolemy's theory fit the data perfectly. But since no one could come up with laws of motion that were remotely plausible explanations for such a path, it seems puzzling that it took a very long time for people to realize that there was something drastically wrong with the theory.

KISS—Keep It Simple, Stupid—is a good motto for lots of things. Complicated theories and proposals and plans are likely to cause foul-ups. In my experience, people who sacrifice comprehensiveness and

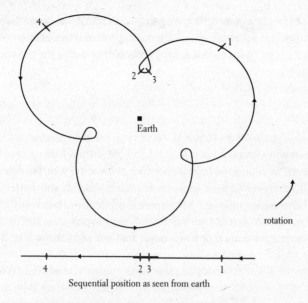

Figure 7. Ptolemy's epicycles to explain the movement of Mars around the earth.

complexity for simplicity are more likely to come up with answers—to something at least, if not to the original question.

Simple theories are to be preferred even when you know they're inadequate to explain all the available evidence. Testing more complicated theories is more labor-intensive, and more likely to lead the investigator down a garden path.

Early in my career, I studied the eating behavior of the obese. I found that their behavior resembled the behavior of rats with lesions to the ventromedial hypothalamus (VMH). Damage to that area of the brain made the rats act as if they were hungry all the time, and they ate enough to become obese. The analogy proved productive, and I was able to show that the feeding behavior of the obese is highly similar to that of rats with VMH lesions. This strongly suggested that obese people are hungry most of the time. I argued that they are attempting to defend a "set point" for weight that is higher for them than for most people.[1] The best evidence for that comes from the fact that the eating behavior of obese people who are not trying to lose weight is the same as that of normal-weight people, whereas the eating behavior of normal-weight people who are trying to lose weight resembles that of obese people who are trying to lose weight.[2]

Experts in the field of eating behavior and obesity told me that the facts couldn't be fully explained by the simple hypothesis of defense of a set point for weight. True enough. But most of the people telling me that didn't really learn much about obesity, whereas people exploring simple hypotheses about obesity have learned a lot.

What's sensible in science is also likely to be sensible in business and other professions. The KISS principle is explicit policy for some highly successful companies and is recommended by many business consultants.

McKinsey & Company instructs its business consultants to keep hypotheses as simple as possible at first, and to allow complications only as they are forced on them.

People who offer advice to start-up companies insist on keeping it simple at first: Release products quickly to get feedback rather than obsessively create the best possible product; target markets where it's possible to maximize early profits rather than aim for a broad range of markets; don't demand complete knowledge of markets or any other aspect of business before acting; keep business models presented to potential investors as simple as possible.

As they say at Google: "Done is better than perfect."

Overly complicated ways of trying to solve a problem are sometimes called Rube Goldberg machines, Goldberg being the fellow who used to draw hilariously convoluted ways of solving a simple problem. For the all-time most spectacular Rube Goldberg machine, click on www .youtube.com/watch?v=qybUFnY7Y8w.

The part of Occam's razor enjoining multiple hypotheses doesn't fully apply to practitioners such as doctors. The more hypotheses the better when we're trying to decide which explanations are in contention and how they can best be examined. I don't want my doctor to entertain only the most plausible hypothesis. I want my doctor to pursue every hypothesis with some reasonable chance of being correct, as well as the possibility that two or more hypotheses are necessary to explain my symptoms. Even for medical diagnosis, though, some rules of parsimony apply. Medical schools teach their students to employ simple and inexpensive diagnostic procedures prior to more complicated and expensive ones and to pursue the most likely possibilities first. ("Think horses, not zebras.")

Reductionism

An issue central to many philosophical and scientific debates concerns *reductionism*, a principle that at first blush resembles Occam's razor. A reductionist explanation is one holding that some seemingly complex phenomenon or system is nothing more than the sum of its parts. Reductionist explanations sometimes go further and maintain that the parts themselves are best understood at some level of complexity simpler than, or lower than, the phenomenon or system itself. That position denies the possibility of *emergence*—in which phenomena come into being that are not explicable solely by invoking processes at a simpler, more basic level. The example par excellence of emergence is consciousness. It has properties that don't exist at the level of the physical, chemical, and electrical events that underlie it (and, so far at least, are not explainable at that level).

If you really can get away with reductionism in either of the senses above, you justifiably win. But the people who study phenomena at a given level are naturally going to be opponents of people who try to dismiss events as mere *epiphenomena*—events secondary to the underlying events and lacking true causal significance.

Some scientists believe that macroeconomics (the aggregate behavior and decision making of the economy as a whole) is fully explained by microeconomics (choices that individuals make). Other scientists believe that microeconomics is fully explained by psychology. Still other scientists believe psychological phenomena can be fully explained by physiological processes, or undoubtedly will be at some future time. And so on. Physiological processes are fully explained by cellular biology, which is fully explained by molecular biology, which is fully explained by chemistry, which is fully explained by the quantum theory of the electromagnetic force, which is fully explained by particle physics. Of course no one proposes that degree of reductionism. But at least some scientists have endorsed one or more of the individual reductions in that chain.

Many reductionist efforts are useful. The principle of parsimony requires us to explain phenomena at the simplest level possible and to add complications only as they become necessary. And the effort to explain things at a level one down in the hierarchy can be useful even if the ultimate conclusion is that there are indeed emergent properties that prevent a full explanation in terms of simpler underlying processes.

But one person's simplification is another person's simple-mindedness. Scientists from other fields are continually trying to explain phenomena in my field of psychology by asserting that they are "nothing but" the operation of factors at some lower level of complexity.

I'll describe two examples of reductionism for psychological events that seem to me to be misguided and off base. Full disclosure: recall that I'm a psychologist!

A decade or so ago a new editor of the prestigious *Science* magazine announced that under his regime the magazine would not accept papers in psychology that failed to show pictures of the brain. This reflected his opinion that psychological phenomena could always be explained at the neural level, or at least that advances in our knowledge of psychological phenomena require at least partial understanding of the brain mechanisms underlying them. Few psychologists, or neuroscientists for that matter, would accept the idea that we're at a stage where purely psychological explanations of psychological phenomena should be considered useless or inadequate. The editor's insistence on physiological reductionism was at best premature.

A much more consequential example of what the philosopher Daniel Dennett calls "greedy reductionism" is the policy formulated a decade or so ago by the head of the National Institute of Mental Health (NIMH) of refusing to support basic research in the behavioral sciences.

The NIMH continues to support basic research in neuroscience and genetics, reflecting the director's highly controversial view that mental illness originates in physiological processes and can be understood primarily or even solely in terms of such processes, rather than as part of a loop among environmental events, mental representations, and biological processes.

Despite the $25 billion that has been spent annually on basic neuroscience research by the National Institutes of Health and the $10 billion spent on basic genetic research, neither type of research has produced new treatments for mental illness. There have been no major advances in the treatment of schizophrenia in fifty years or in the treatment of depression in twenty years.[3]

In contrast, there are many examples of effective treatments for mental illness resulting from basic research in behavioral science, and many more interventions that improve mental health and life satisfaction for normal individuals who would not be considered mentally ill.

We can start with the fact that the theory behind Alcoholics Anonymous, according to its cofounder, came from adopting William James's theories of the role of religion in banishing despair and helplessness.

The best diagnostic procedure available for assessing the likelihood that a person hospitalized for a suicide attempt will make another attempt is called the Implicit Association Test.[4] This measure was originally devised by social psychologists to assess a person's tacit, unrecognized attitudes toward various objects, events, and categories of people. A person whose tacit associations regarding the self are closer to concepts related to death than to life is likely to make a second attempt. Neither the person's self-report, physician's judgment, or any psychiatric test does as good a job at predicting a second attempt.

The most effective treatment for phobias derives from basic research on animal and human learning.

The best available intervention for psychological trauma, discussed in Chapter 10, derives from basic research in social psychology.

And many other examples could be cited.

Finally, behavioral science has been critical in establishing the ineffectiveness, or actual damaging effects, of mental health interventions invented by non–behavioral scientists.

Know Your Own Strength

We don't recognize how easy it is to generate hypotheses about the world. If we did, we'd generate fewer of them, or at least hold them more tentatively. We sprout causal theories in abundance when we learn of a correlation, and we readily find causal explanations for the failure of the world to confirm our hypotheses.

We don't realize how easy it is for us to explain away evidence that would seem on the surface to contradict our hypotheses. And we fail to generate tests of a hypothesis that could falsify the hypothesis if in fact the hypothesis is wrong. This is one type of confirmation bias.

Scientists make all of these mistakes: they sometimes generate hypotheses too readily, they may fail to recognize how easy it is to explain away contrary evidence, and they may not search for procedures that could falsify their hypotheses. Some of the more interesting and important controversies in science involve accusations of unconstrained theorizing, overly facile explanations for apparently contradictory evidence, and failure to recognize opportunities to falsify hypotheses.

An American psychologist once wrote to Freud describing experiments he believed supported Freud's theory of repression. Freud wrote back to him saying that he would have ignored the experiments if they had found "evidence" contradicting his theory; therefore he was obliged to ignore any experimental evidence allegedly supporting it. To his psychoanalytic colleagues, he sniffed *"ganz Amerikanisch"* (completely American).

Freud's put-down seems odd inasmuch as Freud was a committed and highly successful experimenter himself when researching questions of neurology and hypnosis. But Freud's philosophy of science regarding psychoanalysis was that his interpretation of what his patients told him was the royal road to truth. Anyone who disagreed with these interpretations was simply making a grievous error—which he often made clear to any student or colleague who had the temerity to disagree with him.

The scientific community can't accept a claim that only a single individual's judgment counts as evidence. If a theory includes the proviso that only its progenitor (or his acolytes) can assess its truth, then the theory exists to that extent outside of science.

Freud's certainty and dogmatism are the sure signs of someone who is on shaky epistemic ground. And shaky ground is what many, if not most, psychologists and philosophers of science now believe to have been under Freud's feet most of the time.

However, Freud's work gave rise to many hypotheses that *are* testable by normal scientific means, and some of these have received strong support (and not just from Americans!). The notion discussed in Chapter 3 that the unconscious is a preperceiver is one such hypothesis. The evidence is by now overwhelming that people register countless stimuli simultaneously, only some of which get referred to the conscious mind for consideration, and that such nonconscious stimuli can markedly affect behavior. Research strongly supports other psychoanalytic theories. These include the concept of transference—by which feelings about parents or other important individuals that are formed in childhood are transferred more or less intact to other individuals later in life[5]—and sublimation, by which feelings of anger or sexual desire that are unacceptable to a person are channeled into unthreatening activities such as artistic creation.[6]

In the hands of many of its adherents, psychoanalytic theory lacks sufficient *constraints*. For Freud, and for many of his followers, anything goes. If I say that the patient has an "Oedipus complex" (a desire to have sex with his mother), who is to say that's baloney? And on what grounds? "Oedipus shmedipus," as the Jewish mother said, "as long as he loves his mother."

Freud's theory of the psychosexual stages—oral, anal, phallic, latent, and genital—included the assertion that arrested development at one of the early stages was possible and would have a major impact on behavior. Toddlers who withheld their feces instead of making for Mama would be stingy and obsessive-compulsive in adulthood. Freud would never have thought it worthwhile to try to find support for such hypotheses outside of the consulting room. And I very much doubt he would have been successful had he tried.

We would say today that one of the chief ways that psychoanalysts derived their hypotheses was by applying the representativeness heuristic, matching cause and consequence on the basis of their perceived similarity.

The psychoanalytic theorist Bruno Bettelheim deduced that the reason the princess in the fairy tale dislikes the frog is that its "tacky, clammy" feel is connected to children's feelings about sex organs. Who says children don't like their sex organs? (And tacky, clammy? Well . . . never mind.) And what's to keep me from saying that the princess dislikes the frog because the bumps on it remind her of pimples, which she dreads having? Or that she's a nervous Nellie who's startled by the frog's rapid movements?

The concept of the pleasure principle guided Freud's understanding of human nature until the 1920s. Life centered on satisfying the demands of the id for satisfaction of bodily needs, sex, and the discharge of anger. Dreams were usually about wish fulfillment.

But the driving motives of wish fulfillment and the id's desires for life-satisfying gratifications seemed to be contradicted by the need of some Great War trauma victims to keep returning to thoughts about the disastrous events they had encountered. Freud also began to notice that children in their play sometimes fantasized the death of loved ones. Patients who were dealing with painful memories that had previously been repressed kept returning to them obsessively and without seeking resolution. And therapists regularly encountered masochists—people deliberately seeking out pain.

Clearly all these people were not being motivated by the pleasure principle. So there must be some drive opposed to it. Freud labeled this drive the "death instinct"—the desire to return to an inorganic state.

The role of the representativeness heuristic in this hypothesis seems pretty clear. People's main goal in life is the pursuit of pleasure, but sometimes they seem to be seeking the opposite. Therefore there is a drive toward personal extinction. Facile and utterly untestable.

My favorite example of the role of the representativeness heuristic in generating psychoanalytic hypotheses comes from reactions to a paper published in the *American Journal of Psychiatry* by Jules Masserman, at the time the president of the American Psychiatric Association. The burden of the paper, which was intended as a joke, was that ingrown toenails are symbols of masculine aspirations and intrauterine fantasies. To Masserman's chagrin, the journal was flooded by admiring commentary on his perspicacity.[7]

Theories more venerable and better supported by evidence than psychoanalytic theory also have problems with constraints, confirmation, and falsification.

Evolutionary theory has generated thousands of testable and confirmed (or just as frequently disconfirmed) hypotheses about the adaptive nature of the characteristics of organisms. Why are female animals of some species faithful to a single male and the females of other species promiscuous? Perhaps many mates increase the likelihood of reproduction in some species and not in others. Indeed, that turns out to be the case.

Why do some butterflies wear flashy clothes? Explanation: to attract mates. Evidence: male butterflies that have their colors toned down by researchers don't do well in the mating department. Why should the viceroy butterfly mimic nearly perfectly the appearance of the monarch butterfly? Because the monarch is poisonous to most vertebrates and it's to the viceroy's advantage. An animal only needs to fall ill once after eating a monarch in order to avoid ever after anything that resembles a monarch.

But the adaptationist perspective is subject to considerable abuse, and not just by armchair evolutionists.

A construct popular with both cognitive scientists and evolution theorists is the notion of "mental modules"—cognitive structures that evolution has developed for us that guide our ability to deal with some aspect of the world. Mental modules are relatively independent of other mental states and processes and owe little to learning. The clearest example of a mental module is language. No one today would attempt to explain human language as a purely learned phenomenon. The evidence for some degree of prewiring for language is overwhelming: human languages are all similar at some deep level, they are learned at about the same age by people in all cultures, and they are served by specific areas in the brain.

But module explanations by evolutionary theorists are too readily invoked. See a behavior and posit an evolved module for it. There are no obvious constraints on such explanations. They're as easy and unconstrained as are many psychoanalytic explanations.

In addition to the overly facile nature of many evolutionary hypotheses and their violation of Occam's razor, many such hypotheses are not testable by any means currently available. We aren't obligated to pay attention to theories that are untestable. Which isn't to say we're not allowed to believe untestable theories—just that we need to recognize their weakness compared to theories that are. I can believe anything I want about the world, but you have to reckon with it only if I provide evidence for it or an air-tight logical derivation.

The field of psychology affords many examples of too-easy theorizing. Reinforcement learning theory taught us a great deal about the conditions that favor acquisition and "extinction" of learned responses such as a rat's pressing a lever to get food. The theory guided important applications such as treatment of phobias and machine learning procedures. But theorists in that tradition who attempt explanations of complex human behavior in terms of presumed reinforcements sometimes make the same mistakes as many psychoanalytic and evolutionary theorists. Little Oscar does well in school because he was reinforced for conscientious behavior when he was a child, or because other people modeled conscientious behavior for him. How do we know? Because he is now so conscientious in school and does so well there. How else could he have become so conscientious other than through reinforcement for conscientious behavior or behaving like models he has observed being rewarded for such behavior? Hypotheses like that are not merely too easy and unconstrained but circular and not falsifiable by current methods.

Economists of the "rational choice" persuasion sometimes exhibit the same lack of constraint and circular reasoning as psychoanalytic, evolutionary, and learning theorists. All choices are rational because the individual wouldn't have made the choice if he hadn't thought it was in his best interests. We know the person thought it was in his best interests because that's the choice the person made. The near-religious insistence that human choices are always rational leads such economists to make claims that are simultaneously untestable and tautological. The Nobel Prize–winning economist Gary Becker maintained that an individual who chooses to begin a career of drug addiction has to be considered rational if the individual's chief goal in life is to satisfy a need for instant gratification. Facile, irrefutable, and circular. If drug addiction can be "explained" as rational behavior by a rational choice theorist, the theory is bankrupt in that person's hands. All choices are known in advance to be rational, so nothing can be learned about the rationality of any given choice.

But of course my critique isn't limited to scientists. Mea culpa and so are you. Many of the theories we come up with in everyday life are utterly lacking in constraints. They're cheap and lazy, tested if at all by searching only for confirmatory evidence, and too readily salvaged in the face of contradictory evidence.

Judith, a talented young chemist who we thought was surely headed for a distinguished career in science because of her energy and intelligence, has left the field to become a social worker. She must have a fear of success. Too easy to generate that theory and too easy to apply it. And what could convince us that fear of success was not involved?

Bill, mild-mannered neighbor, erupted in rage toward his child at the big-box store. He must have an angry and cruel streak that we hadn't previously seen. The representativeness heuristic, the fundamental attribution error, and the belief in the "law" of small numbers aid and abet one another in producing such theories willy-nilly.

Once generated, evidence that should be considered as disconfirming the hypothesis can be explained away too easily. I have a theory that start-ups supported by large numbers of small investors, even when little information about the company is available, are destined to be highly successful. This applies to the newly founded Bamboozl.com, so it's going to have great success. Bamboozl goes bust, but I'm going to be able to come up with any number of reasons for its failure. Management was not as talented as I had thought. The competition moved much faster than could have been predicted.

I believe that announcement of a cutback of "quantitative easing" by the Federal Reserve will result in fear in the equity markets, causing a drop in stock values. The Fed announces a slowdown of quantitative easing and the markets go up. Because of . . . you name it.

Jennifer, disorganized in her private life, would never make a good newspaper editor, a job that requires meeting deadlines and simultaneously juggling information obtained from Internet sources, assigning tasks to copy editors, and so on. Lo and behold, she turns out to be an excellent editor. The mentoring she got from her predecessor early on must have saved her from the consequences of her fundamentally chaotic temperament.

I'm not saying we shouldn't come up with hypotheses like the above, just that recognition of the ease with which we generate them, and the facility with which we can explain away contradictory evidence, should make us cautious about believing them.

The problem is that we don't recognize our own strength as theorists.

Discussion of theory testing leads us to the question of just what kinds of theories can be falsified and what kinds of evidence could serve to do so.

Falsifiability

If the facts don't fit the theory, change the facts.   —Albert Einstein

No experiment should be believed until it has been confirmed by theory.
                                    —Arthur S. Eddington, astrophysicist

"It's an empirical question" is a statement that ought to end far more conversations than it does.

Deductive reasoning follows logical rules, producing conclusions that can't be refuted if the premises are correct. But most knowledge is obtained not by logic but by collecting evidence. Philosophers call conclusions that are reached by empirical means a form of "defeasible reasoning." That essentially means "defeatable" reasoning. If you look for evidence that plausibly would support your hypothesis, and it does, you may have a reasonable belief. If the data don't support your hypothesis, then you either have to find another way to support the hypothesis or hold on to it with appropriate tentativeness. Or, as Einstein said, show that the "facts" are mistaken.

If someone makes a theoretical claim but can't tell us what kind of evidence would count against it, we should be especially wary of that person's claim. As often as not, the person's simply telling you what ideology or religion has revealed. He's operating in the prophetic mode rather than the empirical tradition.

The falsifiability principle is now enshrined in law in several states as a criterion for teaching something purporting to be science. If it's not falsifiable, it's not science and can't be taught. This is intended primarily to rule out the teaching of creation "science." A typical creationist claim might be, "The human eye is vastly too complicated to have come about by such a cumbersome and laborious process as evolution." The appropriate answer to that proposition is, "Who says?" Such claims are not falsifiable.

The falsifiability requirement makes me slightly nervous, though, because I'm not sure the theory of evolution is falsifiable either. Darwin believed it was. He wrote, "If it could be demonstrated that any complex organ existed, which could not possibly have been formed by numerous, successive, slight modifications, my theory would absolutely break down. But I can find no such case."

And no one has. Or can. If the creationist says that such and such an organ could not have evolved, the evolutionist can only say, "Yes it could." Not very convincing. And there is no way at present to test such claims empirically.

Nevertheless, the theory of evolution wins out over any other theory of the origin of life—of which there are only two others, namely God and seeding by extraterrestrials. Evolution theory triumphs, not because it's falsifiable and has yet to be falsified, but because (a) it's highly plausible, (b) it accounts satisfactorily for countless thousands of diverse and otherwise apparently unrelated facts, (c) it generates hypotheses that are testable, and (d) as the great geneticist Theodosius Dobzhansky said, "Nothing in biology makes sense except in the light of evolution."

The evolutionary hypothesis and the God hypothesis are of course not incompatible. "God works in mysterious ways his wonders to perform." Evolution is actually one of the less mysterious ways an all-powerful being might have chosen to kick-start life and keep it running all the way to us.

Dobzhansky, incidentally, was a religious man. Francis Collins, the leader of the Human Genome Project and the current director of the National Institutes of Health and (obviously) a believer in evolution, is an evangelical Christian. Collins would never pretend that his belief in evolution is of the same kind as his belief in God—which he would be the first to acknowledge is not falsifiable.

Popper and Poppycock

The Austro-British philosopher of science and London School of Economics professor Karl Popper promulgated the view that science proceeds solely by conjecture and falsification of the conjecture or failure to falsify it. Popper maintained that induction is unreliable. In his view we don't (or shouldn't) believe propositions simply because they're supported by evidence from which we induce that the propositions are correct. "All swans are white" was supported by millions of sightings of swans that were white and were never any other color. Oops.

Australian swans are black. Hypotheses can only be disconfirmed, not confirmed.

Popper's injunction is logically correct. No amount of sightings of white swans can establish the truth of the generalization that all swans are white. There is an asymmetry: empirical generalizations can be refuted but can't be proved true because they always rest on inductive evidence that could be refuted at any moment by an exception.

Though correct, Popper's contention is pragmatically useless. We have to act in the world, and falsification is only a small part of the process of generating knowledge to guide our actions. Science advances mostly via induction from facts that support a theory.[8] You have a theory based on deduction from some other theory, or from induction based on observation of available evidence, or from an inspired hunch. You then generate tests of that theory. If they support the theory, you conclude that it's more likely that the theory is correct than it would be in the absence of such evidence. If they don't support the theory, you reduce your confidence in the theory and either look for other tests or put the theory on hold.

Falsification is important in science, for sure. Some facts are powerful enough that they're sufficient to utterly disabuse us of some hypothesis. Observation of chimpanzees that remained inactive and apparently asleep while being operated on after being given curare led to the hypothesis that curare obliterates consciousness. That theory went out the window when the first human was operated on under curare and reported, I assume with expletives, that he had been awake the whole time and felt the surgeon's every excruciating maneuver. The hypothesis that the moon is made of green cheese was destroyed by Neil Armstrong in 1969.

Once you know the knockdown fact, the theory is kaput. (For the time being. Many a theory has been knocked down only to rise again in a modified version.) But mostly, research is a slog through findings that support or contradict the theory to one degree or another.

The glittering prizes in science don't go to the people who falsified someone else's theory, or even one of their own—though their research may have that incidental effect. Rather, the laurels are for scientists who have made predictions based on some novel theory and demonstrated that there are important facts that support the theory and are difficult to explain in the absence of the theory.

Scientists are much more likely to think they accept Popper's anti-inductive stance than philosophers of science are to endorse it. The ones I know think it's utterly wrong. Science advances mostly by induction.

Popper, incidentally, criticized psychoanalytic theory as unfalsifiable and insisted it could therefore be ignored. He was quite mistaken in that. I pointed out earlier that many aspects of the theory are indeed falsifiable, and some have in fact been falsified. The central claims of psychoanalytic theory about therapeutic principles have been, if not refuted, at least shown to be dubious. There is no good evidence that people get better by virtue of dredging up buried memories and working through them with the therapist. And certainly it's the case that psychotherapeutic practices owing nothing to psychoanalytic concepts have been shown to be more effective.

I was told by an eminent philosopher of science that Popper was actually quite ignorant of psychoanalytic theory. He knew only what he picked up in café conversations.

What about Einstein's outrageous comment that facts have to change if they don't support a theory? Many interpretations of the comment are possible, but the one I prefer is that we're allowed to continue to believe a satisfying theory for which there's good support, even though there are facts that are inconsistent with the theory. If the theory is good enough, the "facts" will eventually be overturned. Eddington's quip makes the coordinate point: we're on shaky ground if we believe an alleged fact when there's no plausible theory that should lead us to accept it.

Adherence to Eddington's rule could have spared my field of social psychology a great embarrassment. Its most venerable journal published a wildly implausible claim about extrasensory perception. An investigator asked participants to predict what statement a computer would select at random from a prepared list over a large number of trials. Participants allegedly could predict the behavior of the computer accurately at a level beyond what would be achieved by random guesses. The claim was therefore for paranormal foretelling of future events produced by a machine that could not foretell the events itself. The claim is dismissible on its face; no evidence could support such a theory. Several people with time on their hands tried to replicate the findings and couldn't.

The Hocs: Ad and Post

We have many techniques that allow us to ignore evidence that on its face would seem to contradict our predictions. One of the dodges has to do with dubiously legitimate fixes to a hypothesis. *Ad hoc* postulates are amendments to a theory that don't follow directly from the theory and serve no purpose other than to keep the theory propped up. Ad hoc means literally "to this." (Ad hoc committees are subcommittees of the whole set up to deal with a specific issue.)

Recall from Chapter 14 Aristotle's invention of the property of "levity." This was an ad hoc amendment to the theory that an object's "property" of gravity causes it to fall to earth. Levity was postulated to handle the fact that some things float in water instead of sinking. The concept of levity is a special-purpose fix to Aristotle's theory of gravity, intended to handle a problem that would otherwise wreck the theory. It doesn't follow from the basic theory in any principled way. The theory itself was what I call "placebic." Nothing has actually been explained. The French playwright Molière derides such explanations when he has a character attribute a sleeping potion's effect to its "dormative virtues."

Ptolemy's epicycles were an ad hoc solution to the problem that heavenly bodies did not orbit the earth in the perfect circles that were presumed by his contemporaries to be the necessary pattern of motion.

Einstein's postulation of the cosmological constant, noted in Chapter 14, was a special-purpose fix to the theory of general relativity. It was postulated just to account for the "fact" that the universe was in a steady state. Oops. It isn't in a steady state.

An astronomer has come up with an ad hoc theory to account for the failure of Mercury to orbit the sun in the way demanded by Newton's theory. The astronomer simply posited that the sun's center of gravity shifts from its center to the surface—when and only when the planet in question is Mercury. A desperate (and deliberately hilarious) move to salvage a theory by a special-purpose postulate.

Ad hoc theories are typically also *post hoc*—literally "after this"—meaning a contrivance after the data are in to explain what was not predicted in advance. Post hoc explanations are all too easy to come up with once an anomaly is spotted. "Well, I said I was certain Joan would win the spelling contest, but I couldn't have known that she would be upset by having done badly on a math test the morning of the contest." "Yes, I said Charlie would fail as manager because of his social insensitivity, but I couldn't have guessed that he would marry a woman who would trim the rough edges off him."

In my first years as an academic, I habitually made confident predictions about how someone would function as a department chair or journal editor. When the predictions failed—as they did about as often as not—I had no trouble explaining why any particular prediction was off base. This spared me the necessity of recalibrating my theories about what leads to being a success in a particular role. I'm pleased to be able to say that I now make my predictions with far less certainty than before. Or at least I try to keep my predictions to myself. This saves me considerable embarrassment.

To this point I've tacitly adopted the lay view that scientific investigation and theory building are cut-and-dried procedures following clear rules about hypothesis generation, the search for evidence, and acceptance or rejection of the hypothesis. For better or worse, this is far from the case, as we'll see in the next chapter.

Summing Up

Explanations should be kept simple. They should call on as few concepts as possible, defined as simply as possible. Effects that are the same should be explained by the same cause.

Reductionism in the service of simplicity is a virtue; reductionism for its own sake can be a vice. Events should be explained at the most basic level possible. Unfortunately, there are probably no good rules that can tell us whether an effect is an epiphenomenon lacking causal significance versus a phenomenon emerging from interactions among simpler events and having properties not explainable by those events.

We don't realize how easy it is for us to generate plausible theories. The representativeness heuristic is a particularly fertile source of explanations: we are too inclined to assume that we have a causal explanation for an event if we can point to an event that resembles it. Once generated, hypotheses are given more credence than they deserve because we don't realize we could have generated many different hypotheses with as little effort and knowledge.

Our approach to hypothesis testing is flawed in that we're inclined to search only for evidence that would tend to confirm a theory while failing to search for evidence that would tend to disconfirm it. Moreover, when confronted with apparently disconfirming evidence we're all too skillful at explaining it away.

A theorist who can't specify what kind of evidence would be disconfirmatory should be distrusted. Theories that can't be falsified can be believed, but with the recognition that they're being taken on faith.

Falsifiability of a theory is only one virtue; confirmability is even more important. Contra Karl Popper, science—and the theories that guide our daily lives—change mostly by generating supporting evidence, not by discovering falsifying evidence.

We should be suspicious of theoretical contrivances that are proposed merely to handle apparently disconfirmatory evidence but are not intrinsic to the theory. Ad hoc, post hoc fixes to theories have to be suspect because they are too easy to generate and too transparently opportunistic.

16. Keeping It Real

> There is nothing new to be discovered in physics now. All that remains is
> more and more precise measurement.
> —William Thomson, Lord Kelvin, discoverer of the correct value
> of the temperature of absolute zero, in an address in 1900 to the
> British Association for the Advancement of Science

"Arational" (or nonrational or quasi-rational) practices in science occur
alongside—even in opposition to—the linear, rational textbook version
of scientific progress. Sometimes scientists abandon generally accepted
theories and devote themselves to other theories that are not well supported by available evidence. Their adoption of the new theory is initially
as much a matter of faith as of logic or data.

Scientific theories are sometimes traceable to particular worldviews
that differ across academic fields, between ideologies, or from one culture to another. The different theories sometimes literally conflict with
one another.

The arational aspects of science may have contributed to the rejection
of the concept of objective truth by some people who describe themselves as deconstructionists or postmodernists. What defense is possible
against such nihilism? What can be said to people who assert that "reality"
is mere socially constructed fiction?

Paradigm Shifts

Five years after Lord Kelvin's pronouncement about the boring future
of physics, Einstein published his paper on special relativity. Relativity

theory literally replaced Isaac Newton's mechanics—the laws describing motion and force that had stood unchallenged for two centuries. Einstein's theory was not a mere new development in physics. It heralded a new physics.

Fifty years after Einstein's paper was published, the philosopher and sociologist of science Thomas Kuhn shook the scientific community by announcing in his book *The Structure of Scientific Revolutions* that science doesn't always consist of an earnest slog through theory followed by collection of data followed by adjustment of theory. Rather, revolutions are the customary way that science makes its greatest advances.

The old theory gets creaky, anomalies slowly pile up, and someone has a bright idea that, sooner or later, ends up overthrowing the old theory—or at least rendering it much less relevant and interesting. The new theory typically doesn't account for all the phenomena that the old theory does, and its new contentions are at first supported by data that are underwhelming at best. Often the new theory isn't concerned with explaining established facts at all, but only with predicting new ones.

Kuhn's analysis was upsetting to scientists in part because it introduced an element of seeming irrationality into the concept of scientific progress. Scientists jump ship not so much because the old theory is inadequate or because new data have come in. Rather, a *paradigm shift* occurs because a new idea has come along that is more satisfying in some respects than the old idea, and the scientific program it suggests is more exciting. Scientists seek "low-hanging fruit"—startling findings suggested by the new theory that couldn't be explained by the old theory— that are ripe for the picking.

Often the new theoretical approaches lead nowhere in particular, even though large numbers of scientists are pursuing them. But some new paradigms do break through and replace older views, seemingly overnight.

The field of psychology offers a particularly clear example of the rapid rise of a new paradigm and the near-simultaneous abandonment of an old one.

Psychology from early in the twentieth century till roughly the late 1960s was dominated by reinforcement learning theories. Ivan Pavlov showed that once an animal had learned that a particular arbitrary stimulus signaled a reinforcement of some kind, that stimulus would elicit the same reaction as the reinforcing agent itself. A bell that preceded the introduction of meat would come to produce the same salivary reaction as the meat itself. B. F. Skinner showed that if a given behavior was reinforced by some desirable stimulus, the behavior would be performed whenever the organism wanted the reinforcement. Rats learn to press a lever if that results in food being delivered. Psychologists produced thousands of experiments testing hypotheses derived from one or another principle suggested by Pavlovian and Skinnerian theories.

During the heyday of learning theory, psychologists reached the conclusion that much of human behavior is the result of modeling. I see Jane do something for which she gets a "positive reinforcement." So I learn to do the same thing to get that reinforcement. Or I see her do something that gets her punished, so I learn to avoid that behavior. "Vicarious reinforcement theory" was both obvious and hard to test in a rigorous way, except by hothouse experiments showing that children sometimes imitate other people in the short term. Hit a doll and the child may imitate that. But that doesn't show that chronically aggressive adults got that way by observing other people get rewarded for aggressive behavior.

Among scientifically minded psychologists it was de rigueur to have a reinforcement-learning theory interpretation of every psychological phenomenon, whether it involved the behavior of animals or humans. Scientists who offered different interpretations of the evidence were ignored or worse.

An Achilles' heel of reinforcement theory stems from the fact that it's incrementalist in nature. A light comes on and a shock follows a short time later. The animal slowly learns that the light predicts a shock. Or the animal presses a lever that produces food and the animal gradually learns that lever pressing is its meal ticket.

But phenomena began cropping up in which the animal learned almost instantaneously the connection between two stimuli. For example, an experimenter might periodically deliver an electric shock to a rat shortly after a buzzer sounded. The rat would begin to show fear (indicated by, for example, crouching or defecating) whenever the buzzer sounded. But if a light preceded the buzzer and there was no shock, the rat would show substantially less fear—on the very first trial when the light was introduced. On the next trial there might be virtually no fear expressed at all. This suggested to many people that some types of learning could best be understood as the result of some fairly sophisticated causal thinking on the part of the rat.[1]

Around the same time the temporal puzzles were discovered, Martin Seligman delivered an extremely serious blow to one of the most central tenets of traditional learning theory, namely that you could pair any arbitrary stimulus with any other arbitrary stimulus and an animal would learn that association.[2] Seligman showed that the arbitrariness dictum was hopelessly wrong. Recall from Chapter 8 that associations the animal was not "prepared" to learn would not be learned. Dogs can readily learn to go to the right if a light appears on the right rather than the left, but not if it appears on top rather than on the bottom. Pigeons will starve to death while a learning theorist tries to teach them that *not* pecking at a light will produce a food pellet.

The failures of learning theory to account for the extremely rapid learning of some connections and the impossibility of learning other connections were not initially seen as the body blows that they were. The danger to learning theory came not from these anomalies but from seemingly unrelated work on cognitive processes, including memory, the influence of schemas on visual perception and interpretation of events, and causal reasoning.

Many psychologists began to see that the really exciting phenomena to be examined had to do with thinking rather than learning. Almost overnight hundreds of investigators began studying the operations of the mind, and study of learning processes came to a virtual halt.

Learning theory was not so much disproved as ignored. In retrospect, it can be seen that the program of research had become what the philosopher of science Imre Lakatos termed a "degenerative research paradigm"—one that is no longer producing interesting findings. Just more and more about less and less.

The new opportunities were in the field of cognition (and later in the field of cognitive neuroscience). Within very few years virtually no one was studying learning, and few cognitive scientists deigned to pay attention to learning-theory interpretations of their findings.

As in science, great changes in technology, industry, and commerce are often due to revolution rather than evolution. The steam engine is invented, resulting in the replacement of wool by cotton as the main fabric used for clothing in many parts of the world. Trains are invented, resulting in the deregionalizing of manufacturing. Mass production of goods in factories arrives, ending time-immemorial manufacturing techniques. Within a brief period of time the invention of the Internet changed . . . everything.

One difference between paradigmatic changes in science and those in technology and business practices is that the old paradigm often as not hangs around in science. Cognitive science didn't replace all learning theory findings, or even the explanations behind the findings. Rather, it just established a body of work that couldn't have been produced within the learning theory framework.

Science and Culture

Bertrand Russell once observed that scientists studying the problem-solving behavior of animals saw in their experimental subjects the national characteristics of the scientists themselves. The pragmatic Americans and the theoretically inclined Germans had very different understandings of what was happening.

> Animals studied by Americans rush about frantically, with an incredible display of hustle and pep, and at last achieve the desired result by chance. Animals observed by Germans sit still and think, and at last evolve the solution out of their inner consciousness.

Ouch! Any psychologist knows there was more than a grain of truth in Russell's lampoon. Indeed, the groundwork for the cognitive revolution was laid by Western Europeans, especially Germans, who worked primarily on perception and thinking rather than learning. American soil was pretty barren for cognitive theory, and work on thought would undoubtedly have come along much later if not for prodding by Europeans. It's no accident that social psychology, which was founded by Europeans, was never "behaviorized" in the first place.

In addition to having to acknowledge the arational aspects of paradigm shifts, scientists have had to come to grips with the fact that cultural beliefs can profoundly influence scientific theories.

The Greeks believed in the stability of the universe, and scientists from Aristotle to Einstein were in thrall to this commitment. The Chinese, in contrast, were confident that the world was constantly changing. Chinese attention to context led to their correct understanding of acoustics, magnetism, and gravity.

Continental social scientists shake their heads in exasperation with what they call the rigid "methodological individualism" of American social scientists and their inability to see the relevance or even the existence of larger social structures and of the zeitgeist. The major advances in thinking about societies and organizations have primarily continental rather than Anglo-Saxon roots.

Western primatologists could see no social interaction among chimpanzees more complicated than the behavior that a pair of chimps exhibited toward each other until Japanese primatologists showed the very complicated nature of chimpanzee politics.

Even the preferred forms of reasoning differ across cultures. Logic is foundational for Western thought, dialecticism for East Asian thought. The two types of thinking can produce literally contradictory results.

The rapid and incompletely justified nature of shifts in scientific theories, together with recognition of the role of culture in affecting scientific views, contradicted the picture of science as an enterprise of pure rationality operating in the light of unshakable facts. These deviations may have contributed to a thoroughly antiscientific approach to reality that began to gain steam in the late twentieth century.

Reality as a Text

After we came out of the church, we [Samuel Johnson and his biographer James Boswell] stood talking for some time together of Bishop Berkeley's ingenious sophistry to prove the nonexistence of matter, and that everything in the universe is merely ideal. I [Boswell] observed, that though we are satisfied his doctrine is not true, it is impossible to refute it. I never shall forget the alacrity with which Johnson answered, striking his foot with mighty force against a large stone, till he rebounded from it—"I refute it *thus.*"
—James Boswell, *The Life of Samuel Johnson*

Not everyone today seems to be as readily convinced of the reality of reality as Johnson.

Recall the umpire from Chapter 1 who denied any reality to the concepts of strikes and balls other than his labeling of them as such. Many people who call themselves postmodernists or deconstructionists would endorse that umpire's view.

In Jacques Derrida's phrase: *"Il n'y a pas de hors-texte."* (There is nothing outside of the text.) People with such orientations sometimes deny that there is any "there" there at all. "Reality" is merely a construction, and nothing exists other than our interpretation of it. The fact that interpretations of some aspect of the world can be widely or even universally shared is irrelevant. Such agreement only indicates that there are shared "social constructions." One of my favorite phrases from this movement is that there are no facts—only "regimes of truth."

This extreme subjectivist view drifted over to America from France in the 1970s. The general idea behind deconstructionism is that texts can be dismantled to show the ideological leanings, values, and arbitrary perspectives that underlie all inferences about the world, including assertions posing as facts about nature.

An anthropologist of my acquaintance was asked by a student at my university how anthropologists deal with the problem of reliability concerning characterizations of the beliefs and behavior of people in other cultures. In other words, what to do about the sometimes varying interpretations of different anthropologists? She replied, "The problem doesn't arise because what we anthropologists do is interpret what we see. Different people are expected to have different interpretations because of their different assumptions and viewpoints."

This answer scandalized my student—and me. If you're doing science, agreement is everything. If observers can't agree about whether a given phenomenon exists, then scientific interpretation can't even get launched. What you have is a mess.

But my mistake was in thinking that cultural anthropologists necessarily regard themselves as scientists. Early on in my work on cultural psychology I tried to make contact with cultural anthropologists. I wanted to learn from them, and I expected they would be interested in my empirical work on cultural differences in thought and behavior. I was shocked to discover that most of the people defining themselves as cultural anthropologists had no desire to talk to me and no use for my data. They were not about to "privilege" (their term) my evidence over their interpretations.

To my astonishment, postmodernist nihilism made strong headway in academic fields ranging from literary studies to history to sociology. How strong? An acquaintance told me about asking a student whether she thought the laws of physics were mere arbitrary assertions about nature. "Yes," she assured her questioner. "Well, when you're up in an airplane you figure any old laws of physics could keep it in the air?" "Absolutely," she replied. A survey of students at a major university by the philosopher and political scientist James Flynn found that most believed that modern science is merely one point of view.[3] Those poor students came by their opinion honestly. It was encouraged by the sort of thing they had been told in many of their humanities and social science courses. One might think that professors in those fields were merely amusing themselves or perhaps trying to stimulate thought on the part of their students. But consider the tale of the physicist and the postmodernists.

In 1996, Alan Sokal, a physics professor at New York University, sent a manuscript to *Social Text*, a journal with a proudly postmodern stance and an editorial roster including some quite famous academics. Sokal's article, titled "Transgressing the Boundaries: Towards a Transformative Hermeneutics of Quantum Gravity," tested just how much nonsense such a journal was willing to swallow. The article, saturated with postmodern jargon, announced that "an external world whose properties are independent of any individual human being" was "dogma imposed by the long post-Enlightenment hegemony over the Western intellectual outlook." Because scientific research is "inherently theory laden and self-referential," it "cannot assert a privileged epistemological status with respect to counterhegemonic narratives emanating from dissident or marginalized communities." Quantum gravity was pronounced a mere social construction.

Sokal's article was accepted without peer review. On the day of publication of his article in *Social Text*, Sokal revealed in the journal *Lingua Franca* that the article was a pseudoscientific hoax. The editors of *Social Text* responded that the article's "status as parody does not alter, substantially, our interest in the piece, itself, as a symptomatic document."

George Orwell said that some things are so stupid that only intellectuals believe them. But to be fair, no one actually believes that reality is merely a text, though many people undoubtedly think they believe it. Or did. Postmodernism is gradually fading from the North American academic scene. It dissipated long ago in France, where, as my French anthropologist friend Dan Sperber said, "it never even had the prestige of being French!"

Should you find yourself in a conversation with a postmodernist, and I can't wholeheartedly recommend you do, try the following. Ask whether the balance on the person's credit card statement is a mere social construction. Or ask whether he thinks power differentials in society are merely a matter of interpretation or whether they have some basis in reality.

I have to admit, incidentally, that postmodernist concerns have produced some research related to power, ethnicity, and gender that seems valid and important. The anthropologist Ann Stoler, for example, has done very interesting research on the shaky and sometimes hilarious criteria used by the Dutch in colonial Indonesia to determine who was and was not "white." Nothing so straightforward as the American rule that anyone with a "single drop" of African blood was a Negro, which of course was a social construction without any remote basis in physical reality. Stoler's work is of substantial interest to historians, to anthropologists, and to psychologists interested in how people categorize the world and how people's motivations influence their understanding of the world.

What I find particularly ironic about postmodernists is that they asserted without evidence that interpretations of reality are always just that, and did so in the complete absence of knowledge about the findings by psychologists that support a contention only slightly less radical on its face than postmodernists' views. One of the greatest accomplishments of psychologists is the demonstration of the philosopher's dictum that everything from the perception of motion to understanding of the workings of our own minds is an inference. Nothing in the world is known as directly or infallibly as intuition tells us it is.

But the fact that everything is an inference doesn't mean that any inference is as defensible as another. Should you find yourself at the zoo with a postmodernist, don't let him get away with telling you that your belief that the large animal with the trunk and tusks is an elephant is a mere inference—because it could be a mouse with a glandular condition.

Summing Up

Science is based not only on evidence and well-justified theories—faith and hunches may cause scientists to ignore established scientific hypotheses and agreed-upon facts. Several years ago, the literary agent John Brockman asked scores of scientists and public figures to tell him about something they believed that they couldn't prove—and he published their responses in a book.[4] In many instances, an individual's most important work was guided by hypotheses that could never be proved. As laypeople we have no choice but to do the same.

The paradigms that underlie a given body of scientific work, as well as those that form the basis for technologies, industries, and commercial enterprises, are subject to change without notice. These changes are often initially "underdetermined" by the evidence. Sometimes the new paradigm exists in uneasy partnership with the old, and sometimes it utterly replaces the old.

Different cultural practices and beliefs can produce different scientific theories, paradigms, and even forms of reasoning. The same is true for different business practices.

Quasi-rational practices by scientists, and cultural influences on belief systems and reasoning patterns, may have encouraged postmodernists and deconstructionists to press the view that there are no facts, only socially agreed-upon interpretations of reality. They clearly don't live their lives as if they believed this, but they nevertheless expended a colossal amount of university teaching and "research" effort promulgating these nihilistic views. Did these teachings contribute to the rejection of scientific findings in favor of personal prejudices so common today?

Conclusion: The Tools of the Lay Scientist

This book has brought you some bad news and some good news.

The bad news is that our beliefs about many important aspects of the world are often sorely mistaken, and the ways in which we acquire them are often fundamentally flawed.

Our conviction that we know the world directly, by unmediated perception of facts, is what philosophers call "naive realism." Every belief about every aspect of the world is based on countless inferences we make via mental processes we can't observe. We're dependent on innumerable schemas and heuristics to categorize accurately even the simplest objects and events.

We frequently fail to see the role of context in producing the behavior of humans and even of physical objects. We're often oblivious to the role played by the social influences that drive our judgments and guide our behavior.

Countless stimuli affect our beliefs and behavior without our knowledge, sometimes even without our awareness of their existence.

Our belief that we know what goes on in our heads is far wide of the mark. When we can correctly identify the mental processes that produced some judgment or solved some problem, we do so not by observing those processes but by applying theories about those processes. And those theories are often wrong.

We're overly influenced by anecdotal evidence. This problem is amplified by our failure to understand the importance of having lots of information relevant to the judgment at hand. We operate as if we thought the law of large numbers also applied to small numbers. We're particularly blind to the possibility that our evidence is insufficient when it comes to making some of the most important judgments, namely about the characteristics of other people.

We have great difficulty correctly identifying relationships between even highly important events. If we think there's a relationship, we're likely to see it even if it isn't there. If we think there's not likely to be a relationship, we frequently fail to detect it even when it's quite strong.

We generate theories about the world with abandon, having little understanding that the ease with which we do so is no indication a given theory is correct. In particular, we're profligate causal theorists. Given an effect, we readily, even automatically and without reflection, come up with a theory about its cause. Even when we think to test the theory, we're flawed as intuitive scientists. We tend to look exclusively for confirming evidence while failing to look for equally probative evidence that might serve to discredit the theory. When we're forced to confront disconfirming evidence, we're gifted at explaining it away, being unaware of how easy it is for us to generate ad hoc defenses of our original theory.

The bottom line for all this: our beliefs are often badly mistaken, we're way too confident about our ability to acquire new knowledge that accurately characterizes the world, and our behavior often fails to advance our interests and those of people we care about.

The good news is the flip side of the bad news. You already knew you were fallible before you read this book. You now know much more about what produces your failings and how to compensate for them. This knowledge will help you perceive the world more accurately and behave more sensibly. What you've read also serves as a weapon to guard against the flawed assertions of others—friends and acquaintances as well as people in the media.

You'll often apply the concepts and rules you've learned automatically, even without awareness that you're applying them. And that's going to be increasingly true over time.

Use a new tool in this book a few times and you'll frequently have it when you need it. You're not going to forget the law of large numbers and its implications for the amount of evidence needed, and every time you apply the law you're going to be more likely to use it in the future—in an ever-widening range of events. You're not going to forget the admonition to pay more attention to social context in order to explain your behavior and that of others. On the contrary, you'll be getting constant feedback showing you that you've understood some situation better than you would have in the past and that reinforcement is going to result in your applying the concept more and more frequently in the future. You're going to have the sunk cost and opportunity cost concepts available to you for the rest of your life.

So you're a better scientist in your everyday life than when you began this book. But I don't want to oversell just how much you're going to be able to change the way you think. I violate most of the principles in this book frequently and many of them constantly. Some of our psychological tendencies are just very deeply rooted, and they're not going to be extirpated by learning some new principles intended to reduce their untoward effects. But I know these tendencies can be modified, and their damage limited, by virtue of knowing about them and how to combat them.

You're also a better consumer and media critic now than when you began this book. Let's consider a couple of reports and one letter to the editor that I read in highly respected newspapers while I was drafting this final chapter.

- *The New York Times* reported that married couples who had big weddings had longer lasting, more satisfying marriages than couples who had smaller weddings.[1] But I'm betting that you wouldn't start encouraging your friends to crank out more wedding invitations. I'm hoping it would have occurred to you that people who have big weddings are, on average, older, better off financially, have known each other longer, and are possibly more in love than people who have smaller weddings. All these factors are correlated with marital happiness. We have learned precisely nothing from the finding that there is a correlation between size of wedding and marriage satisfaction.
- The Associated Press reported highway safety data for a large number of 2011 model autos. Findings included the fact that death rates per million cars for Subaru Legacy sedans and Toyota Highlander hybrid SUVs, among other autos, were vastly lower than death rates for Chevrolet Silverado 1500 pickups and Jeep Patriot SUVs, for example. I'm hoping that had you read that article you would have considered that the death rate per car is

a less accurate measure of safety than the death rate per mile, since average number of miles driven undoubtedly differs substantially across vehicle types. More important to consider are the characteristics of the typical drivers of the vehicles. Which type of car is most likely to be driven by the proverbial little old lady from Pasadena or a Westchester County, New York, soccer mom? Which is most likely to be driven by hell-raising young Texas cowboys or pampered California teenagers?

- *The Wall Street Journal* published a letter in 2012 from an MIT climate scientist and others maintaining that global warming is minimal and apparently ceasing, citing as evidence the fact that there had been no increase in global temperatures since 1998. I would hope it might have occurred to you to think about what the standard deviation might be for temperature changes from one year to another. They're quite large, actually. Moreover, as for any partially random process, there are surprisingly many long runs. Temperature change, like many phenomena, doesn't move in a straight line, but rather in fits and starts. And, in fact, 2014 turned out to be the hottest year on record. (There were a couple of other reasons to be dubious about the letter. Signers of the letter included a geneticist, a spaceship designer, and someone described as a former astronaut and U.S. senator, suggesting that the bottom of the expertise barrel was being scraped. And the letter compared the firing of a journal editor, allegedly for writing an article doubting climate change, to the imprisonment and execution of Soviet scientists who doubted Lysenko's genetic views. No kidding.)

So in many cases you'll be able to rebut, or at least have solid reasons for doubting, claims by acquaintances and the media that you might have accepted before. But, more often than in the past, you're going to be alert to the fact that you just don't have the tools to test a given claim. Few of us can critique claims such as "stents are better than coronary artery grafting for most plugged aorta problems," or "amino acids from crashed comets could have laid the groundwork for life on earth," or "the oil reserves on the American continental shelf exceed those of Saudi

Arabia's." We're all lay scientists at best with respect to almost all information we get about almost all domains. So normally you'll have to turn to other sources. That would be putative experts in the field pertinent to your concerns. What's the proper stance toward the experts in a given field—assuming you can find them?

Here are the philosopher Bertrand Russell's "mild propositions" about how to deal with expert opinion.

- When the experts are agreed, the opposite opinion cannot be held to be certain.
- When they are not agreed, no opinion can be regarded as certain by a nonexpert.
- When they all hold that no sufficient grounds for a positive opinion exist, the ordinary man would do well to suspend his judgment.

Mild propositions indeed. Too mild, maybe?

Many years ago I attended a psychology department talk by someone who billed himself as a computer scientist. Not many people used that job title in those days. The speaker began by announcing, "I am going to deal with the question of what it might mean to humans' conceptions of themselves if one day computers could beat any international chess master, write a better novel or symphony than any human, and solve fundamental questions about the nature of the world that have stumped the greatest intellects throughout history."

His next utterance produced an audible gasp from the audience. "I want to make two things clear at the outset. First, I don't know whether computers will ever be able to do those things. Second, I'm the only person in the room with a right to an opinion on the question."

The second sentence has rung in my ears ever since that day. The speaker shocked me into the habit of subjecting other people's claims—and my own—to the expertise test. You constantly hear people express firm opinions about some matter for which there may be—in fact you know there to be—expert opinion available. Does the person have a right to claim expertise, like the computer scientist I heard decades ago? Does the person believe his opinion is based on that of experts? Does the person know what range of opinions there is among the experts? Does the person even know whether there *are* experts? Does the person *care* whether there are experts?

Scientists certainly care about whether there are experts. They often make progress by questioning the received wisdom of experts. My career illustrates that. It's been one long series of discoveries that experts, usually including me at the outset of my research, can be mistaken. Here are some of the dozens of cases where I've found the experts to be dead wrong.

- Many obese people are not overeating, as the experts (and I) believed, but rather defending a set point for fat tissue.
- People don't have introspective access to their mental processes as cognitive psychologists (including me) believed. Rather, when they're right about what went on in their heads it's because they happen to have a correct theory about how they arrived at a given judgment or solved a particular problem. Often, however, such theories are mistaken.
- Like most people studying statistical reasoning, I was confident that the teaching of statistical principles could have only minimal effects on people's reasoning in everyday life. Thank goodness I was wrong, and this book is due in part to that discovery.
- Economists and reinforcement-theory psychologists have long believed that incentives—usually of the monetary sort—are the best way to change behavior. But monetary incentives are often useless or worse, and there are many other less expensive and less coercive ways to change behavior.
- For the better part of a century, experts in the field of intelligence agreed that intelligence is essentially one thing, namely IQ as measured by standard tests; that it's little influenced by environmental factors; and that differences in IQ between blacks and whites are due in part to genes. All of that is wrong.[2]

I have some expertise that allowed me to confront expert opinion in all these areas. But unfortunately my expertise is limited to the small number of fields I've worked in. I'm pretty much just a lay scientist with respect to everything else. And that's true for all of us. So how to regard the experts in the fields we need to know something about?

I'd go further than Bertrand Russell. It's not just that you shouldn't be certain of an opinion opposite to that of the experts when the experts are agreed. Rather, it would seem unwise not to simply accept their opinion—unless you have solid grounds for believing that you have some alternate expertise that allows you to doubt the general consensus. It's foolish to assume that our ignorance, or the views of an entertainment celebrity delivered on a talk show, are better guides to the truth than the experts' knowledge.

Of course, it can be very hard to find out what the consensus of experts is on many matters. Indeed, in the name of "balance," the media often do their best to confuse you as to whether there is a consensus. If they have a presumed expert giving her view on some issue, they find another "expert" with a different view. I often see this balancing act when I know for sure that the strong consensus of experts holds to one view over the other. The near-universal consensus among climate experts is that change is occurring, due at least in part to human activities. Yet it's been reported that the Fox News president, Roger Ailes, has standing orders that anyone presenting this view must be rebutted by someone who denies the correctness of the consensus.

So it's easy to be misled by the media, whether motivated by political goals or more often by a misguided insistence on balance, into believing there's a significant division of expert opinion, and therefore that it's reasonable to pick among various positions. But, believe me, you can always find someone with a PhD to support any crank opinion. Evolution? Hooey. Alien visits to our planet? Without doubt. Vaccinations cause autism? Absolutely. Megadoses of vitamin C combat the common cold? Darn tootin'.

It's getting easier to know what the consensus of experts is on a given subject. Fortunately, in fields where it's important for us to have accurate knowledge, such as health and education, there are reputable websites such as Mayo Clinic and the What Works Clearinghouse that make the job easier. But the Internet is not a panacea. I can assure you that anything having to do with gender differences in behavior, and some things having to do with gender differences in biology, should be viewed with a beady eye.

See what you think about my suggestions for how to approach the question of expert opinion about matters that are important to you or to society as a whole.

1. Try to find out whether there is such a thing as expertise about the question. There is no expertise about astrology.
2. If there is such a thing as expertise, try to find out whether there is a consensus among the experts.
3. If there is a consensus, then the stronger that consensus seems to be, the less choice you have about whether to accept it.

Winston Churchill said, "Democracy is the worst form of government except for all those others that have been tried." Experts are the worst people to trust except for all those other people whose views you might consult.

And bear in mind that I'm an expert on the question of the expertise of experts!

Notes

INTRODUCTION
1. Gould, *The Panda's Thumb*.
2. Nisbett, "Hunger, Obesity and the Ventromedial Hypothalamus."
3. Polanyi, *Personal Knowledge*.
4. Nisbett, *The Geography of Thought*.
5. Lehman et al., "The Effects of Graduate Training on Reasoning"; Lehman, Darrin, and Nisbett, "A Longitudinal Study of the Effects of Undergraduate Education on Reasoning"; Morris and Nisbett, "Tools of the Trade."
6. Larrick, Morgan, and Nisbett, "Teaching the Use of Cost-Benefit Reasoning in Everyday Life"; Larrick, Nisbett, and Morgan, "Who Uses the Cost-Benefit Rules of Choice? Implications for the Normative Status of Microeconomic Theory"; Nisbett et al., "Teaching Reasoning"; Nisbett et al., "Improving Inductive Inference" in Kahneman, Slovic, and Tversky, *Judgment Under Uncertainty*; Nisbett et al., "The Use of Statistical Heuristics in Everyday Reasoning."

1. EVERYTHING'S AN INFERENCE
1. Shepard, *Mind Sights: Original Visual Illusions, Ambiguities, and Other Anomalies*.
2. Higgins, Rholes, and Jones, "Category Accessibility and Impression Formation."
3. Bargh, "Automaticity in Social Psychology."
4. Cesario, Plaks, and Higgins, "Automatic Social Behavior as Motivated Preparation to Interact."
5. Darley and Gross, "A Hypothesis-Confirming Bias in Labeling Effects."
6. Meyer and Schvaneveldt, "Facilitation in Recognizing Pairs of Words: Evidence of a Dependence Between Retrieval Operations."
7. Ross and Ward, "Naive Realism in Everyday Life: Implications for Social Conflict and Misunderstanding."
8. Jung et al., "Female Hurricanes Are Deadlier Than Male Hurricanes."
9. Alter, *Drunk Tank Pink*.
10. Berman, Jonides, and Kaplan, "The Cognitive Benefits of Interacting with Nature"; Lichtenfield et al., "Fertile Green: Green Facilitates Creative Performance"; Mehta and Zhu, "Blue or Red? Exploring the Effect of Color on Cognitive Task Performances."
11. Alter, *Drunk Tank Pink*.

12. Berger, Meredith, and Wheeler, "Contextual Priming: Where People Vote Affects How They Vote."
13. Rigdon et al., "Minimal Social Cues in the Dictator Game."
14. Song and Schwarz, "If It's Hard to Read, It's Hard to Do."
15. Lee and Schwarz, "Bidirectionality, Mediation, and Moderation of Metaphorical Effects: The Embodiment of Social Suspicion and Fishy Smells."
16. Alter and Oppenheimer, "Predicting Stock Price Fluctuations Using Processing Fluency."
17. Danziger, Levav, and Avnaim-Pesso, "Extraneous Factors in Judicial Decisions."
18. Williams and Bargh, "Experiencing Physical Warmth Influences Personal Warmth."
19. Dutton and Aron, "Some Evidence for Heightened Sexual Attraction Under Conditions of High Anxiety."
20. Levin and Gaeth, "Framing of Attribute Information Before and After Consuming the Product."
21. McNeil et al., "On the Elicitation of Preferences for Alternative Therapies."
22. Daniel Kahneman, *Thinking, Fast and Slow.*
23. Tversky and Kahneman, "Extensional Versus Intuitive Reasoning: The Conjunction Fallacy in Probability Judgment."
24. Tversky and Kahneman, "Judgment Under Uncertainty: Heuristics and Biases."
25. Gilovich, Vallone, and Tversky, "The Hot Hand in Basketball: On the Misperception of Random Sequences."

2. THE POWER OF THE SITUATION

 1. Jones and Harris, "The Attribution of Attitudes."
 2. Darley and Latané, "Bystander Intervention in Emergencies: Diffusion of Responsibility."
 3. Darley and Batson, "From Jerusalem to Jericho: A Study of Situational and Dispositional Variables in Helping Behavior."
 4. Pietromonaco and Nisbett, "Swimming Upstream Against the Fundamental Attribution Error: Subjects' Weak Generalizations from the Darley and Batson Study."
 5. Humphrey, "How Work Roles Influence Perception: Structural-Cognitive Processes and Organizational Behavior."
 6. Triplett, "The Dynamogenic Factors in Pacemaking and Competition."
 7. Brown, Eicher, and Petrie, "The Importance of Peer Group ('Crowd') Affiliation in Adolescence."
 8. Kremer and Levy, "Peer Effects and Alcohol Use Among College Students."
 9. Prentice and Miller, "Pluralistic Ignorance and Alcohol Use on Campus."
10. Liu et al., "Findings from the 2008 Administration of the College Senior Survey (CSS): National Aggregates."
11. Sanchez-Burks, "Performance in Intercultural Interactions at Work: Cross-Cultural Differences in Responses to Behavioral Mirroring."
12. Goethals and Reckman, "The Perception of Consistency in Attitudes."
13. Goethals, Cooper, and Naficy, "Role of Foreseen, Foreseeable, and Unforeseeable Behavioral Consequences in the Arousal of Cognitive Dissonance."
14. Nisbett et al., "Behavior as Seen by the Actor and as Seen by the Observer."

15. Ibid.
16. Nisbett, *The Geography of Thought*; Nisbett et al., "Culture and Systems of Thought: Holistic Vs. Analytic Cognition."
17. Masuda et al., "Placing the Face in Context: Cultural Differences in the Perception of Facial Emotion."
18. Masuda and Nisbett, "Attending Holistically vs. Analytically: Comparing the Context Sensitivity of Japanese and Americans."
19. Cha and Nam, "A Test of Kelley's Cube Theory of Attribution: A Cross-Cultural Replication of McArthur's Study."
20. Choi and Nisbett, "Situational Salience and Cultural Differences in the Correspondence Bias and in the Actor-Observer Bias."
21. Nisbett, *The Geography of Thought*.

3. THE RATIONAL UNCONSCIOUS

1. Nisbett and Wilson, "Telling More Than We Can Know: Verbal Reports on Mental Processes."
2. Zajonc, "The Attitudinal Effects of Mere Exposure."
3. Bargh and Pietromonaco, "Automatic Information Processing and Social Perception: The Influence of Trait Information Presented Outside of Conscious Awareness on Impression Formation."
4. Karremans, Stroebe, and Claus, "Beyond Vicary's Fantasies: The Impact of Subliminal Priming and Brand Choice."
5. Chartrand et al., "Nonconscious Goals and Consumer Choice."
6. Berger and Fitzsimons, "Dogs on the Street, Pumas on Your Feet."
7. Buss, *The Murderer Next Door: Why the Mind Is Designed to Kill*.
8. Wilson and Schooler, "Thinking Too Much: Introspection Can Reduce the Quality of Preferences and Decisions."
9. Dijksterhuis and Nordgren, "A Theory of Unconscious Thought."
10. The interpretation that I (and the authors) prefer for the art poster, jam, and apartment studies has been questioned. I side with the authors, but the following references will get you in touch with the available evidence pro and con concerning the possibility that unconscious pondering of alternatives can result in superior choices: Aczel et al., "Unconscious Intuition or Conscious Analysis: Critical Questions for the Deliberation-Without-Attention Paradigm"; Calvillo and Penaloza, "Are Complex Decisions Better Left to the Unconscious?"; Dijksterhuis, "Think Different: The Merits of Unconscious Thought in Preference Development and Decision Making"; Dijksterhuis and Nordgren, "A Theory of Unconscious Thought"; A. Dijksterhuis et al., "On Making the Right Choice: The Deliberation-Without-Attention Effect"; Gonzalo et al., "'Save Angels Perhaps': A Critical Examination of Unconscious Thought Theory and the Deliberation-Without-Attention Effect"; Strick et al., "A Meta-Analysis on Unconscious Thought Effects."
11. Lewicki et al., "Nonconscious Acquisition of Information."
12. Klarreich, "Unheralded Mathematician Bridges the Prime Gap."
13. Ghiselin, ed. *The Creative Process*.
14. Maier, "Reasoning in Humans II: The Solution of a Problem and Its Appearance in Consciousness."

15. Kim, "Naked Self-Interest? Why the Legal Profession Resists Gatekeeping"; O'Brien, Sommers, and Ellsworth, "Ask and What Shall Ye Receive? A Guide for Using and Interpreting What Jurors Tell Us"; Thompson, Fong, and Rosenhan, "Inadmissible Evidence and Juror Verdicts."

4. SHOULD YOU THINK LIKE AN ECONOMIST?
1. Dunn, Aknin, and Norton, "Spending Money on Others Promotes Happiness."
2. Borgonovi, "Doing Well by Doing Good: The Relationship Between Formal Volunteering and Self-Reported Health and Happiness."
3. Heckman, "Skill Formation and the Economics of Investing in Disadvantaged Children"; Knudsen et al., "Economic, Neurobiological, and Behavioral Perspectives on Building America's Future Workforce."
4. Sunstein, "The Stunning Triumph of Cost-Benefit Analysis."
5. Appelbaum, "As U.S. Agencies Put More Value on a Life, Businesses Fret."
6. NBC News, "How to Value Life? EPA Devalues Its Estimate."
7. Appelbaum, "As U.S. Agencies Put More Value on a Life, Businesses Fret."
8. Kingsbury, "The Value of a Human Life: $129,000."
9. Desvousges et al., "Measuring Non-Use Damages Using Contingent Valuation: An Experimental Evaluation of Accuracy."
10. Hardin, "The Tragedy of the Commons."

5. SPILT MILK AND FREE LUNCH
1. Larrick, Morgan, and Nisbett, "Teaching the Use of Cost-Benefit Reasoning in Everyday Life"; Larrick, Nisbett, and Morgan, "Who Uses the Cost-Benefit Rules of Choice? Implications for the Normative Status of Microeconomic Theory." These papers report this and all the remaining findings in this section.
2. Larrick, Nisbett, and Morgan, "Who Uses the Cost-Benefit Rules of Choice? Implications for the Normative Status of Microeconomic Theory."
3. Larrick, Morgan, and Nisbett, "Teaching the Use of Cost-Benefit Reasoning in Everyday Life."

6. FOILING FOIBLES
1. Thaler and Sunstein, *Nudge: Improving Decisions About Health, Wealth, and Happiness.*
2. Kahneman, Knetch, and Thaler, "Experimental Tests of the Endowment Effect and the Coase Theorem."
3. Kahneman, *Thinking, Fast and Slow.*
4. Fryer et al., "Enhancing the Efficacy of Teacher Incentives Through Loss Aversion: A Field Experiment."
5. Kahneman, *Thinking, Fast and Slow.*
6. Samuelson and Zeckhauser, "Status Quo Bias in Decision Making."
7. Thaler and Sunstein, *Nudge: Improving Decisions About Health, Wealth, and Happiness.*

8. Ibid.

9. Investment Company Institute, "401(K) Plans: A 25-Year Retrospective."

10. Thaler and Sunstein, *Nudge: Improving Decisions About Health, Wealth, and Happiness.*

11. Madrian and Shea, "The Power of Suggestion: Inertia in 401(K) Participation and Savings Behavior."

12. Benartzi and Thaler, "Heuristics and Biases in Retirement Savings Behavior."

13. Iyengar and Lepper, "When Choice Is Demotivating: Can One Desire Too Much of a Good Thing?"

14. Thaler and Sunstein, *Nudge: Improving Decisions About Health, Wealth, and Happiness.*

15. Ibid.

16. Schultz et al., "The Constructive, Destructive, and Reconstructive Power of Social Norms."

17. Perkins, Haines, and Rice, "Misperceiving the College Drinking Norm and Related Problems: A Nationwide Study of Exposure to Prevention Information, Perceived Norms and Student Alcohol Misuse"; Prentice and Miller, "Pluralistic Ignorance and Alcohol Use on Campus."

18. Goldstein, Cialdini, and Griskevicius, "A Room with a Viewpoint: Using Social Norms to Motivate Environmental Conservation in Hotels."

19. Lepper, Greene, and Nisbett, "Undermining Children's Intrinsic Interest with Extrinsic Reward: A Test of the Overjustification Hypothesis."

PART III: CODING, COUNTING, CORRELATION, AND CAUSALITY

1. Lehman, Lempert, and Nisbett, "The Effects of Graduate Training on Reasoning: Formal Discipline and Thinking About Everyday Life Events."

7. ODDS AND *N*S

1. Kuncel, Hezlett, and Ones, "A Comprehensive Meta-Analysis of the Predictive Validity of the Graduate Record Examinations: Implications for Graduate Student Selection and Performance."

2. Kunda and Nisbett, "The Psychometrics of Everyday Life."

3. Rein and Rainwater, "How Large Is the Welfare Class?"

4. Kahneman, *Thinking, Fast and Slow.*

8. LINKED UP

1. Smedslund, "The Concept of Correlation in Adults"; Ward and Jenkins, "The Display of Information and the Judgment of Contingency."

2. Zagorsky, "Do You Have to Be Smart to Be Rich? The Impact of IQ on Wealth, Income and Financial Distress."

3. Kuncel, Hezlett, and Ones, "A Comprehensive Meta-Analysis of the Predictive Validity of the Graduate Record Examinations: Implications for Graduate Student Selection and Performance."

4. Schnall et al., "The Relationship Between Religion and Cardiovascular Outcomes and All-Cause Mortality: The Women's Health Initiative Observational Study (Electronic Version)."

5. Arden et al., "Intelligence and Semen Quality Are Positively Correlated."

6. Chapman and Chapman, "Genesis of Popular but Erroneous Diagnostic Observations."

7. Ibid.

8. Seligman, "On the Generality of the Laws of Learning."

9. Jennings, Amabile, and Ross, "Informal Covariation Assessment: Data-Based Vs. Theory-Based Judgments," in Tversky and Kahneman, *Judgment Under Uncertainty.*

10. Valochovic et al., "Examiner Reliability in Dental Radiography."

11. Keel, "How Reliable Are Results from the Semen Analysis?"

12. Lu et al., "Comparison of Three Sperm-Counting Methods for the Determination of Sperm Concentration in Human Semen and Sperm Suspensions."

13. Kunda and Nisbett, "Prediction and the Partial Understanding of the Law of Large Numbers."

14. Ibid.

15. Fong, Krantz, and Nisbett, "The Effects of Statistical Training on Thinking About Everyday Problems."

9. IGNORE THE HiPPO

1. Christian, "The A/B Test: Inside the Technology That's Changing the Rules of Business."

2. Carey, "Academic 'Dream Team' Helped Obama's Effort."

3. Moss, "Nudged to the Produce Aisle by a Look in the Mirror."

4. Ibid.

5. Ibid.

6. Cialdini, *Influence: How and Why People Agree to Things.*

7. Silver, *The Signal and the Noise.*

10. EXPERIMENTS NATURAL AND EXPERIMENTS PROPER

1. See, e.g., McDade et al., "Early Origins of Inflammation: Microbial Exposures in Infancy Predict Lower Levels of C-Reactive Protein in Adulthood."

2. Bisgaard et al., "Reduced Diversity of the Intestinal Microbiota During Infancy Is Associated with Increased Risk of Allergic Disease at School Age."

3. Olszak et al., "Microbial Exposure During Early Life Has Persistent Effects on Natural Killer T Cell Function."

4. Slomski, "Prophylactic Probiotic May Prevent Colic in Newborns."

5. Balistreri, "Does Childhood Antibiotic Use Cause IBD?"

6. Ibid.

7. Ibid.

8. Hamre and Pianta, "Can Instructional and Emotional Support in the First-Grade Classroom Make a Difference for Children at Risk of School Failure?"

9. Kuo and Sullivan, "Aggression and Violence in the Inner City: Effects of Environment via Mental Fatigue."

10. Nisbett, *Intelligence and How to Get It: Why Schools and Cultures Count.*

11. Deming, "Early Childhood Intervention and Life-Cycle Skill Development."
12. Magnuson, Ruhm, and Waldfogel, "How Much Is Too Much? The Influence of Preschool Centers on Children's Social and Cognitive Development."
13. Roberts et al., "Multiple Session Early Psychological Interventions for Prevention of Post-Traumatic Disorder."
14. Wilson, *Redirect: The Surprising New Science of Psychological Change.*
15. Pennebaker, "Putting Stress into Words: Health, Linguistic and Therapeutic Implications."
16. Wilson, *Redirect: The Surprising New Science of Psychological Change.*
17. Ibid.
18. Ibid.
19. Prentice and Miller, "Pluralistic Ignorance and Alcohol Use on Campus."

11. EEKONOMICS

1. Cheney, "National Center on Education and the Economy: New Commission on the Skills of the American Workforce."
2. Heraty, Morley, and McCarthy, "Vocational Education and Training in the Republic of Ireland: Institutional Reform and Policy Developments Since the 1960s."
3. Hanushek, "The Economics of Schooling: Production and Efficiency in Public Schools"; Hoxby, "The Effects of Class Size on Student Achievement: New Evidence from Population Variation"; Jencks et al., *Inequality: A Reassessment of the Effects of Family and Schooling in America.*
4. Krueger, "Experimental Estimates of Education Production Functions."
5. Shin and Chung, "Class Size and Student Achievement in the United States: A Meta-Analysis."
6. Samieri et al., "Olive Oil Consumption, Plasma Oleic Acid, and Stroke Incidence."
7. Fong et al., "Correction of Visual Impairment by Cataract Surgery and Improved Survival in Older Persons."
8. Samieri et al., "Olive Oil Consumption, Plasma Oleic Acid, and Stroke Incidence."
9. Humphrey and Chan, "Postmenopausal Hormone Replacement Therapy and the Primary Prevention of Cardiovascular Disease."
10. Klein, "Vitamin E and the Risk of Prostate Cancer."
11. Offit, *Do You Believe in Magic? The Sense and Nonsense of Alternative Medicine.*
12. Ibid.
13. Lowry, "Caught in a Revolving Door of Unemployment."
14. Kahn, "Our Long-Term Unemployment Challenge (in Charts)."
15. Bertrand and Mullainathan, "Are Emily and Greg More Employable Than Lakisha and Jamal? A Field Experiment on Labor Market Discrimination."
16. Fryer and Levitt, "The Causes and Consequences of Distinctively Black Names."
17. Ibid.
18. Ibid.
19. Ibid.
20. Milkman, Akinola, and Chugh, "Temporal Distance and Discrimination: An Audit Study in Academia." Additional analysis of data provided by Milkman.
21. Levitt and Dubner, *Freakonomics.*
22. Ibid.

23. Ibid.
24. I have discussed the evidence on the importance of the environment for intelligence in Nisbett, *Intelligence and How to Get It*, and in Nisbett et al., "Intelligence: New Findings and Theoretical Developments."
25. Munk, *The Idealist*.
26. Ibid.
27. Mullainathan and Shafir, *Scarcity: Why Having Too Little Means So Much*.
28. Chetty, Friedman, and Rockoff, "Measuring the Impacts of Teachers II: Teacher Value-Added and Student Outcomes in Adulthood."
29. Fryer, "Financial Incentives and Student Achievement: Evidence from Randomized Trials."
30. Fryer et al., "Enhancing the Efficacy of Teacher Incentives Through Loss Aversion: A Field Experiment."
31. Kalev, Dobbin, and Kelley, "Best Practices or Best Guesses? Assessing the Efficacy of Corporate Affirmative Action and Diversity Policies."
32. Ayres, "Fair Driving: Gender and Race Discrimination in Retail Car Negotiations."
33. Zebrowitz, *Reading Faces: Window to the Soul?*

12. DON'T ASK, CAN'T TELL

 1. Strack, Martin, and Stepper, "Inhibiting and Facilitating Conditions of the Human Smile: A Nonobtrusive Test of the Facial Feedback Hypothesis."
 2. Caspi and Elder, "Life Satisfaction in Old Age: Linking Social Psychology and History."
 3. Schwarz and Clore, "Mood, Misattribution, and Judgments of Well-Being: Informative and Directive Functions of Affective States."
 4. Schwarz, Strack, and Mai, "Assimilation-Contrast Effects in Part-Whole Question Sequences: A Conversational Logic Analysis."
 5. Asch, "Studies in the Principles of Judgments and Attitudes."
 6. Ellsworth and Ross, "Public Opinion and Capital Punishment: A Close Examination of the Views of Abolitionists and Retentionists."
 7. Saad, "U.S. Abortion Attitudes Closely Divided."
 8. Ibid.
 9. Weiss and Brown, "Self-Insight Error in the Explanation of Mood."
10. Peng, Nisbett, and Wong, "Validity Problems Comparing Values Across Cultures and Possible Solutions."
11. Schmitt et al., "The Geographic Distribution of Big Five Personality Traits: Patterns and Profiles of Human Self-Description Across 56 Nations."
12. Heine et al., "What's Wrong with Cross-Cultural Comparisons of Subjective Likert Scales?: The Reference Group Effect."
13. Naumann and John, "Are Asian Americans Lower in Conscientiousness and Openness?"
14. College Board, "Student Descriptive Questionnaire."
15. Heine and Lehman, "The Cultural Construction of Self-Enhancement: An Examination of Group-Serving Biases."
16. Heine, *Cultural Psychology*.

17. Straub, "Mind the Gap: On the Appropriate Use of Focus Groups and Usability Testing in Planning and Evaluating Interfaces."

13. LOGIC

1. Morris and Nisbett, "Tools of the Trade: Deductive Reasoning Schemas Taught in Psychology and Philosophy"; Nisbett, *Rules for Reasoning*.
2. Cheng and Holyoak, "Pragmatic Reasoning Schemas"; Cheng et al., "Pragmatic Versus Syntactic Approaches to Training Deductive Reasoning."
3. Cheng and Holyoak, "Pragmatic Reasoning Schemas"; Cheng et al., "Pragmatic Versus Syntactic Approaches to Training Deductive Reasoning."
4. Lehman and Nisbett, "A Longitudinal Study of the Effects of Undergraduate Education on Reasoning."
5. Ibid.

14. DIALECTICAL REASONING

1. Graham, *Later Mohist Logic, Ethics, and Science*.
2. Ibid.
3. Chan, "The Story of Chinese Philosophy"; Disheng, "China's Traditional Mode of Thought and Science: A Critique of the Theory That China's Traditional Thought Was Primitive Thought."
4. Peng, "Naive Dialecticism and Its Effects on Reasoning and Judgment About Contradiction"; Peng and Nisbett, "Culture, Dialectics, and Reasoning About Contradiction"; Peng, Spencer-Rodgers, and Nian, "Naive Dialecticism and the Tao of Chinese Thought."
5. Ji, Su, and Nisbett, "Culture, Change and Prediction."
6. Ji, Zhang, and Guo, "To Buy or to Sell: Cultural Differences in Stock Market Decisions Based on Stock Price Trends."
7. Peng and Nisbett, "Culture, Dialectics, and Reasoning About Contradiction."
8. Ara Norenzayan et al., "Cultural Preferences for Formal Versus Intuitive Reasoning."
9. Norenzayan and Kim, "A Cross-Cultural Comparison of Regulatory Focus and Its Effect on the Logical Consistency of Beliefs."
10. Watanabe, "Styles of Reasoning in Japan and the United States: Logic of Education in Two Cultures."
11. Logan, *The Alphabet Effect*.
12. Flynn, *Asian Americans: Achievement Beyond IQ*.
13. Ibid.
14. Dweck, *Mindset: The New Psychology of Success*.
15. Aronson, Fried, and Good, "Reducing Stereotype Threat and Boosting Academic Achievement of African-American Students: The Role of Conceptions of Intelligence."
16. Basseches, "Dialectical Schemata: A Framework for the Empirical Study of the Development of Dialectical Thinking"; Basseches, *Dialectical Thinking and Adult Development*; Riegel, "Dialectical Operations: The Final Period of Cognitive Development."

17. Grossmann et al., "Aging and Wisdom: Culture Matters"; Grossmann et al., "Reasoning About Social Conflicts Improves into Old Age."
18. Grossmann et al., "Aging and Wisdom: Culture Matters."
19. Grossmann et al., "Reasoning About Social Conflicts Improves into Old Age."

PART VI: KNOWING THE WORLD
1. Stich, ed., *Collected Papers: Knowledge, Rationality, and Morality, 1978–2010.*

15. KISS AND TELL
1. Nisbett, "Hunger, Obesity and the Ventromedial Hypothalamus."
2. Herman and Mack, "Restrained and Unrestrained Eating."
3. Akil et al., "The Future of Psychiatric Research: Genomes and Neural Circuits."
4. Nock et al., "Measuring the Suicidal Mind: Implicit Cognition Predicts Suicidal Behavior."
5. Kraus and Chen, "Striving to Be Known by Significant Others: Automatic Activation of Self-Verification Goals in Relationship Contexts"; Andersen, Glassman, and Chen, "Transference Is Social Perception: The Role of Chronic Accessibility in Significant-Other Representations."
6. Cohen, Kim, and Hudson, "Religion, the Forbidden, and Sublimation"; Hudson and Cohen, "Taboo Desires, Creativity, and Career Choice."
7. Samuel, *Shrink: A Cultural History of Psychoanalysis in America.*
8. Lakatos, *The Methodology of Scientific Research Programmes: Philosophical Papers*, volume 1.

16. KEEPING IT REAL
1. Holyoak, Koh, and Nisbett, "A Theory of Conditioning: Inductive Learning Within Rule-Based Default Hierarchies"; Kamin, "'Attention-Like' Processes in Classical Conditioning."
2. Seligman, "On the Generality of the Laws of Learning."
3. Flynn, *How to Improve Your Mind: Twenty Keys to Unlock the Modern World.*
4. Brockman, *What We Believe but Cannot Prove.*

CONCLUSION: THE TOOLS OF THE LAY SCIENTIST
1. Parker-Pope, "The Decisive Marriage."
2. Nisbett, *Intelligence and How to Get It: Why Schools and Cultures Count.*

Bibliography

Aczel, B., B. Lukacs, J. Komlos, and M.R.F. Aitken. "Unconscious Intuition or Conscious Analysis: Critical Questions for the Deliberation-Without-Attention Paradigm." *Judgment and Decision Making* 6 (2011): 351–58.

Akil, Huda, et al. "The Future of Psychiatric Research: Genomes and Neural Circuits." *Science* 327 (2010): 1580–81.

AlDabal, Laila, and Ahmed S. BaHammam. "Metabolic, Endocrine, and Immune Consequences of Sleep Deprivation." *Open Respiratory Medicine Journal* 5 (2011): 31–43.

Alter, Adam. *Drunk Tank Pink*. New York: Penguin Group, 2013.

Alter, Adam, and Daniel M. Oppenheimer. "Predicting Stock Price Fluctuations Using Processing Fluency." *Proceedings of the National Academy of Science* 103 (2006): 9369–72.

Andersen, Susan M., Noah S. Glassman, and Serena Chen. "Transference Is Social Perception: The Role of Chronic Accessibility in Significant-Other Representations." *Journal of Personality and Social Psychology* 69 (1995): 41–57.

Appelbaum, Binyamin. "As U.S. Agencies Put More Value on a Life, Businesses Fret." *The New York Times*. Published electronically February 16, 2011. http://www.nytimes.com/2011/02/17/business/economy/17regulation.html?pagewanted=all&_r=0.

Arden, Rosalind, L. S. Gottfredson, G. Miller, and A. Pierce. "Intelligence and Semen Quality Are Positively Correlated." *Intelligence* 37 (2008): 277–82.

Aronson, Joshua, Carrie B. Fried, and Catherine Good. "Reducing Stereotype Threat and Boosting Academic Achievement of African-American Students: The Role of Conceptions of Intelligence." *Journal of Experimental Social Psychology* 38 (2002): 113–25.

Asch, S. E. "Studies in the Principles of Judgments and Attitudes: II. Determination of Judgments by Group and by Ego Standards." *Journal of Social Psychology* 12 (1940): 584–88.

Aschbacher, K., et al. "Combination of Caregiving Stress and Hormone Therapy Is Associated with Prolonged Platelet Activation to Acute Stress Among Postmenopausal Women." *Psychosomatic Medicine* 69 (2008): 910–17.

Ayres, Ian. "Fair Driving: Gender and Race Discrimination in Retail Car Negotiations." *Harvard Review* 104 (1991): 817–72.

Balistreri, William F. "Does Childhood Antibiotic Use Cause IBD?" *Medscape Today* (January 2013). http://www.medscape.com/viewarticle/777412.

Bargh, John A. "Automaticity in Social Psychology." In *Social Psychology: Handbook of Basic Principles*, edited by E. T. Higgins and A. W. Kruglanski, 1–40. New York: Guilford, 1996.

Bargh, John A., and Paula Pietromonaco. "Automatic Information Processing and Social Perception: The Influence of Trait Information Presented Outside of Conscious Awareness on Impression Formation." *Journal of Personality and Social Psychology* 43 (1982): 437–49.

Basseches, Michael. "Dialectical Schemata: A Framework for the Empirical Study of the Development of Dialectical Thinking." *Human Development* 23 (1980): 400–21.

———. *Dialectical Thinking and Adult Development*. Norwood, NJ: Ablex, 1984.

Beccuti, Guglielmo, and Silvana Pannain. "Sleep and Obesity." *Current Open Clinical Nutrition and Metabolic Care* 14 (2011): 402–12.

Benartzi, Shlomo, and Richard H. Thaler. "Heuristics and Biases in Retirement Savings Behavior." *Journal of Economic Perspectives* 21 (2007): 81–104.

Berger, Jonah, and Gráinne M. Fitzsimons. "Dogs on the Street, Pumas on Your Feet." *Journal of Marketing Research* 45 (2008): 1–14.

Berger, Jonah, M. Meredith, and S. C. Wheeler. "Contextual Priming: Where People Vote Affects How They Vote." *Proceedings of the National Academy of Science* 105 (2008): 8846–49.

Berman, M. G., J. Jonides, and S. Kaplan. "The Cognitive Benefits of Interacting with Nature." *Psychological Science* 19 (2008): 1207–12.

Bertrand, Marianne, and Sendhil Mullainathan. "Are Emily and Greg More Employable Than Lakisha and Jamal? A Field Experiment on Labor Market Discrimination." National Bureau of Economic Research Working Paper No. 9873, 2003.

Bisgaard, H., N. Li, K. Bonnelykke, B.L.K. Chawes, T. Skov, G. Pauldan-Muller, J. Stokholm, B. Smith, and K. A. Krogfelt. "Reduced Diversity of the Intestinal Microbiota During Infancy Is Associated with Increased Risk of Allergic Disease at School Age." *Journal of Allergy and Clinical Immunology* 128 (2011): 646–52.

Borgonovi, Francesca. "Doing Well by Doing Good: The Relationship Between Formal Volunteering and Self-Reported Health and Happiness." *Social Science and Medicine* 66 (2008): 2321–34.

Brockman, John. *What We Believe but Cannot Prove*. New York: HarperCollins, 2006.

Brown, B. Bradford, Sue Ann Eicher, and Sandra Petrie. "The Importance of Peer Group ('Crowd') Affiliation in Adolescence." *Journal of Adolescence* 9 (1986): 73–96.

Buss, David M. *The Murderer Next Door: Why the Mind Is Designed to Kill*. New York: Penguin, 2006.

Calvillo, D. P., and A. Penaloza. "Are Complex Decisions Better Left to the Unconscious?" *Judgment and Decision Making* 4 (2009): 509–17.

Carey, Benedict. "Academic 'Dream Team' Helped Obama's Effort." *The New York Times*. Published electronically Nov. 13, 2013. http://www.nytimes.com/2012/11/13/health/dream-team-of-behavioral-scientists-advised-obama-campaign.html?pagewanted=all.

Caspi, Avshalom, and Glen H. Elder. "Life Satisfaction in Old Age: Linking Social Psychology and History." *Psychology and Aging* 1 (1986): 18–26.

Cesario, J., J. E. Plaks, and E. T. Higgins. "Automatic Social Behavior as Motivated Preparation to Interact." *Journal of Personality and Social Psychology* 90 (2006): 893–910.

Cha, J-H, and K. D. Nam. "A Test of Kelley's Cube Theory of Attribution: A Cross-Cultural Replication of McArthur's Study." *Korean Social Science Journal* 12 (1985): 151–80.

Chan, W. T. "The Story of Chinese Philosophy." In *The Chinese Mind: Essentials of Chinese Philosophy and Culture*, edited by C. A. Moore. Honolulu: East-West Center Press, 1967.

Chapman, Loren J., and Jean P. Chapman. "Genesis of Popular but Erroneous Diagnostic Observations." *Journal of Abnormal Psychology* 72 (1967): 193–204.

Chartrand, Tanya L., J. Huber, B. Shiv, and R. J. Tanner. "Nonconscious Goals and Consumer Choice." *Journal of Consumer Research* 35 (2008): 189–201.

Cheney, Gretchen. "National Center on Education and the Economy: New Commission on the Skills of the American Workforce." National Center on Education and the Economy, 2006.

Cheng, P. W., and K. J. Holyoak. "Pragmatic Reasoning Schemas." *Cognitive Psychology* 17 (1985): 391–416.

Cheng, P. W., K. J. Holyoak, R. E. Nisbett, and L. Oliver. "Pragmatic Versus Syntactic Approaches to Training Deductive Reasoning." *Cognitive Psychology* 18 (1986): 293–328.

Chetty, Rag, John Friedman, and Jonah Rockoff. "Measuring the Impacts of Teachers II: Teacher Value-Added and Student Outcomes in Adulthood." *American Economic Review* 104 (2014): 2633–79.

Choi, Incheol. "The Cultural Psychology of Surprise: Holistic Theories, Contradiction, and Epistemic Curiosity." PhD thesis, University of Michigan, 1998.

Choi, Incheol, and Richard E. Nisbett. "Situational Salience and Cultural Differences in the Correspondence Bias and in the Actor-Observer Bias." *Personality and Social Psychology Bulletin* 24 (1998): 949–60.

Christian, Brian. "The A/B Test: Inside the Technology That's Changing the Rules of Business." *Wired* (2012). http://www.wired.com/business/2012/04/ff_abtesting/.

Cialdini, Robert B. *Influence: How and Why People Agree to Things*. New York: Quill, 1984.

CNN. "Germ Fighting Tips for a Healthy Baby." http://www.cnn.com/2011/HEALTH/02/02/healthy.baby.parenting/index.html.

Cohen, Dov, Emily Kim, and Nathan W. Hudson. "Religion, the Forbidden, and Sublimation." *Current Directions in Psychological Science* (2014): 1–7.

College Board. "Student Descriptive Questionnaire." Princeton, NJ: Educational Testing Service, 1976–77.

CTV. "Infants' Exposure to Germs Linked to Lower Allergy Risk." http://www.ctvnews.ca/infant-s-exposure-to-germs-linked-to-lower-allergy-risk-1.720556.

Danziger, Shai, J. Levav, and L. Avnaim-Pesso. "Extraneous Factors in Judicial Decisions." *Proceedings of the National Academy of Science* 108 (2011): 6889–92.

Darley, John M., and C. Daniel Batson. "From Jerusalem to Jericho: A Study of Situational and Dispositional Variables in Helping Behavior." *Journal of Personality and Social Psychology* 27 (1973): 100–119.

Darley, John M., and P. H. Gross. "A Hypothesis-Confirming Bias in Labeling Effects." *Journal of Personality and Social Psychology* 44 (1983): 20–33.

Darley, John M., and Bibb Latané. "Bystander Intervention in Emergencies: Diffusion of Responsibility." *Journal of Personality and Social Psychology* 8 (1968): 377–83.

Deming, David. "Early Childhood Intervention and Life-Cycle Skill Development." *American Economic Journal: Applied Economics* (2009): 111–34.

Desvousges, William H., et al. "Measuring Non-Use Damages Using Contingent Valuation: An Experimental Evaluation of Accuracy." In *Research Triangle Institute Monograph 92-1.* Research Triangle Park, NC: Research Triangle Institute, 1992.

Dijksterhuis, Ap. "Think Different: The Merits of Unconscious Thought in Preference Development and Decision Making." *Journal of Personality and Social Psychology* 87 (2004): 586–98.

Dijksterhuis, Ap, M. W. Bos, L. F. Nordgren, and R. B. van Baaren. "On Making the Right Choice: The Deliberation-Without-Attention Effect." *Science* 311 (2006): 1005–1007.

Dijksterhuis, Ap, and Loran F. Nordgren. "A Theory of Unconscious Thought." *Perspectives on Psychological Science* 1 (2006): 95.

Disheng, Y. "China's Traditional Mode of Thought and Science: A Critique of the Theory That China's Traditional Thought Was Primitive Thought." *Chinese Studies in Philosophy* (Winter 1990–91): 43–62.

Dunn, Elizabeth W., Laura B. Aknin, and Michael I. Norton. "Spending Money on Others Promotes Happiness." *Science* 319 (2008): 1687–88.

Dutton, Donald G., and Arthur P. Aron. "Some Evidence for Heightened Sexual Attraction Under Conditions of High Anxiety." *Journal of Personality and Social Psychology* 30 (1974): 510–51.

Dweck, Carol S. *Mindset: The New Psychology of Success.* New York: Random House, 2010.

Ellsworth, Phoebe C., and Lee Ross. "Public Opinion and Capital Punishment: A Close Examination of the Views of Abolitionists and Retentionists." *Crime and Delinquency* 29 (1983): 116–69.

Flynn, James R. *Asian Americans: Achievement Beyond IQ.* Hillsdale, NJ: Lawrence Erlbaum, 1991.

———. *How to Improve Your Mind: Twenty Keys to Unlock the Modern World.* London: Wiley-Blackwell, 2012.

Fong, Calvin S., P. Mitchell, E. Rochtchina, E. T. Teber, T. Hong, and J. J. Wang. "Correction of Visual Impairment by Cataract Surgery and Improved Survival in Older Persons." *Ophthalmology* 120 (2013): 1720–27.

Fong, Geoffrey T., David H. Krantz, and Richard E. Nisbett. "The Effects of Statistical Training on Thinking About Everyday Problems." *Cognitive Psychology* 18 (1986): 253–92.

Fryer, Roland G. "Financial Incentives and Student Achievement: Evidence from Randomized Trials." *Quarterly Journal of Economics* 126 (2011): 1755–98.

Fryer, Roland G., and Steven D. Levitt. "The Causes and Consequences of Distinctively Black Names." *The Quarterly Journal of Economics* 119 (2004): 767–805.

Fryer, Roland G., Steven D. Levitt, John List, and Sally Sadoff. "Enhancing the Efficacy of Teacher Incentives Through Loss Aversion: A Field Experiment." National Bureau of Economic Research Working Paper No. 18237, 2012.

Ghiselin, Brewster, ed. *The Creative Process.* Berkeley and Los Angeles: University of California Press, 1952/1980.

Gilovich, Thomas, Robert Vallone, and Amos Tversky. "The Hot Hand in Basketball:

On the Misperception of Random Sequences." *Cognitive Personality* 17 (1985): 295–314.

Goethals, George R., Joel Cooper, and Anahita Naficy. "Role of Foreseen, Foreseeable, and Unforeseeable Behavioral Consequences in the Arousal of Cognitive Dissonance." *Journal of Personality and Social Psychology* 37 (1979): 1179–85.

Goethals, George R., and Richard F. Reckman. "The Perception of Consistency in Attitudes." *Journal of Experimental Social Psychology* 9 (1973): 491–501.

Goldstein, Noah J., Robert B. Cialdini, and Vladas Griskevicius. "A Room with a Viewpoint: Using Social Norms to Motivate Environmental Conservation in Hotels." *Journal of Consumer Research* 35 (2008): 472–82.

Gonzalo, C., D. G. Lassiter, F. S. Bellezza, and M. J. Lindberg. " 'Save Angels Perhaps': A Critical Examination of Unconscious Thought Theory and the Deliberation-Without-Attention Effect." *Review of General Psychology* 12 (2008): 282–96.

Gould, Stephen J. "The Panda's Thumb." In *The Panda's Thumb*. New York: W. W. Norton, 1980.

Graham, Angus C. *Later Mohist Logic, Ethics, and Science*. Hong Kong: Chinese U, 1978.

Grossmann, Igor, Mayumi Karasawa, Satoko Izumi, Jinkyung Na, Michael E. W. Varnum, Shinobu Kitayama, and Richard E. Nisbett. "Aging and Wisdom: Culture Matters." *Psychological Science* 23 (2012): 1059–66.

Grossmann, Igor, Jinkyung Na, Michael E. W. Varnum, Denise C. Park, Shinobu Kitayama, and Richard E. Nisbett. "Reasoning About Social Conflicts Improves into Old Age." *Proceedings of the National Academy of Sciences* 107 (2010): 7246–50.

Hamre, B. K., and R. C. Pianta. "Can Instructional and Emotional Support in the First-Grade Classroom Make a Difference for Children at Risk of School Failure?" *Child Development* 76 (2005): 949–67.

Hanushek, Eric A. "The Economics of Schooling: Production and Efficiency in Public Schools." *Journal of Economic Literature* 24 (1986): 1141–77.

Hardin, Garrett. "The Tragedy of the Commons." *Science* 162 (1968): 1243–45.

Heckman, James J. "Skill Formation and the Economics of Investing in Disadvantaged Children." *Science* 312 (2006): 1900–1902.

Heine, Steven J. *Cultural Psychology*. New York: W. W. Norton, 2008.

Heine, Steven J., and Darrin R. Lehman. "The Cultural Construction of Self-Enhancement: An Examination of Group-Serving Biases." *Journal of Personality and Social Psychology* 72 (1997): 1268–83.

Heine, Steven J., Darrin R. Lehman, K. Peng, and J. Greenholtz. "What's Wrong with Cross-Cultural Comparisons of Subjective Likert Scales?: The Reference Group Effect." *Journal of Personality and Social Psychology* 82 (2002): 903–18.

Heraty, Noreen, Michael J. Morley, and Alma McCarthy. "Vocational Education and Training in the Republic of Ireland: Institutional Reform and Policy Developments Since the 1960s." *Journal of Vocational Education and Training* 52 (2000): 177–99.

Herman, C. Peter, and Deborah Mack. "Restrained and Unrestrained Eating." *Journal of Personality* 43 (1975): 647–60.

Higgins, E. Tory, W. S. Rholes, and C. R. Jones. "Category Accessibility and Impression Formation." *Journal of Experimental Social Psychology* 13 (1977): 141–54.

Holyoak, Keith J., Kyunghee Koh, and Richard E. Nisbett. "A Theory of Conditioning: Inductive Learning Within Rule-Based Default Hierarchies." *Psychological Review* 96 (1989): 315–40.

Hoxby, Caroline M. "The Effects of Class Size on Student Achievement: New Evidence from Population Variation." *Quarterly Journal of Economics* 115 (2000): 1239–85.

Hudson, Nathan W., and Dov Cohen. "Taboo Desires, Creativity, and Career Choice." Unpublished manuscript, 2014.

Humphrey, Linda L., and Benjamin K. S. Chan. "Postmenopausal Hormone Replacement Therapy and the Primary Prevention of Cardiovascular Disease." *Annals of Internal Medicine* 137 (2002). Published electronically August 20, 2002. http://annals.org/article.aspx?articleid=715575.

Humphrey, Ronald. "How Work Roles Influence Perception: Structural-Cognitive Processes and Organizational Behavior." *American Sociological Review* 50 (1985): 242–52.

Inhelder, B., and J. Piaget. *The Growth of Logical Thinking from Childhood to Adolescence.* New York: Basic Books, 1958.

Investment Company Institute. "401(K) Plans: A 25-Year Retrospective." 2006. http://www.ici.org/pdf/per12-02.pdf.

Iyengar, Sheena S., and Mark R. Lepper. "When Choice Is Demotivating: Can One Desire Too Much of a Good Thing?" *Journal of Personality and Social Psychology* 79 (2000): 995–1006.

Jencks, Christopher, M. Smith, H. Acland, M. J. Bane, D. Cohen, H. Gintis, B. Heyns, and S. Mitchelson. *Inequality: A Reassessment of the Effects of Family and Schooling in America.* New York: Harper and Row, 1972.

Jennings, Dennis, Teresa M. Amabile, and Lee Ross. "Informal Covariation Assessment: Data-Based Vs. Theory-Based Judgments." In *Judgment Under Uncertainty: Heuristics and Biases,* edited by Amos Tversky and Daniel Kahneman. New York: Cambridge University Press, 1980.

Ji, Li-Jun, Yanjie Su, and Richard E. Nisbett. "Culture, Change and Prediction." *Psychological Science* 12 (2001): 450–56.

Ji, Li-Jun, Zhiyong Zhang, and Tieyuan Guo. "To Buy or to Sell: Cultural Differences in Stock Market Decisions Based on Stock Price Trends." *Journal of Behavioral Decision Making* 21 (2008): 399–413.

Jones, Edward E., and Victor A. Harris. "The Attribution of Attitudes." *Journal of Experimental Social Psychology* 3 (1967): 1–24.

Jung, K., S. Shavitt, M. Viswanathan, and J. M. Hilbe. "Female Hurricanes Are Deadlier Than Male Hurricanes." *Proceedings of the National Academy of Science* (2014). Published electronically June 2, 2014.

Kahn, Robert. "Our Long-Term Unemployment Challenge (in Charts)." 2013. http://blogs.cfr.org/kahn/2013/04/17/our-long-term-unemployment-challenge-in-charts/.

Kahneman, Daniel. *Thinking, Fast and Slow.* New York: Farrar, Straus and Giroux, 2011.

Kahneman, Daniel, Jack L. Knetch, and Richard H. Thaler. "Experimental Tests of the Endowment Effect and the Coase Theorem." In *Tastes for Endowment, Identity, and the Emotions,* vol. 3 of *The New Behavioral Economics,* edited by E. L. Khalil, 119–42. International Library of Critical Writings in Economics. Cheltenham, U.K.: Elgar, 2009.

Kalev, Alexandra, Frank Dobbin, and Erin Kelley. "Best Practices or Best Guesses? Assessing the Efficacy of Corporate Affirmative Action and Diversity Policies." *American Sociological Review* 71 (2006): 589–617.

Kamin, Leon J. "'Attention-Like' Processes in Classical Conditioning." In *Miami Symposium on the Prediction of Behavior: Aversive Stimulation*, edited by M. R. Jones. Miami, FL: University of Miami Press, 1968.

Karremans, Johan C., Wolfgang Stroebe, and Jasper Claus. "Beyond Vicary's Fantasies: The Impact of Subliminal Priming and Brand Choice." *Journal of Experimental Social Psychology* 42 (2006): 792–98.

Keel, B. A. "How Reliable Are Results from the Semen Analysis?" *Fertility and Sterility* 82 (2004): 41–44.

Kim, Sung Hui. "Naked Self-Interest? Why the Legal Profession Resists Gatekeeping." *Florida Law Review* 63 (2011): 129–62.

Kingsbury, Kathleen. "The Value of a Human Life: $129,000." *Time*. Published electronically May 20, 2008. http://www.time.com/time/health/article/0,8599,1808049,00.html.

Klarreich, Erica. "Unheralded Mathematician Bridges the Prime Gap." *Quanta Magazine*. May 19, 2013. www.quantamagazine.org/20130519-unheralded-mathematician-bridges-the-prime-gap/.

Klein, E. A. "Vitamin E and the Risk of Prostate Cancer: The Selenium and Vitamin E Cancer Prevention Trial." *Journal of the American Medical Association* 306 (2011). Published electronically October 12, 2011. http://jama.jamanetwork.com/article.aspx?articleid=1104493.

Knudsen, Eric I., J. J. Heckman, J. L. Cameron, and J. P. Shonkoff. "Economic, Neurobiological, and Behavioral Perspectives on Building America's Future Workforce." *Proceedings of the National Academy of Science* 103 (2006): 10155–62.

Kraus, Michael W., and Serena Chen. "Striving to Be Known by Significant Others: Automatic Activation of Self-Verification Goals in Relationship Contexts." *Journal of Personality and Social Psychology* 97 (2009): 58–73.

Kremer, Michael, and Dan M. Levy. "Peer Effects and Alcohol Use Among College Students." National Bureau of Economic Research Working Paper No. 9876, 2003.

Krueger, Alan B. "Experimental Estimates of Education Production Functions." *Quarterly Journal of Economics* 114 (1999): 497–532.

Kuncel, Nathan R., Sarah A. Hezlett, and Deniz S. Ones. "A Comprehensive Meta-Analysis of the Predictive Validity of the Graduate Record Examinations: Implications for Graduate Student Selection and Performance." *Psychological Bulletin* 127 (2001): 162–81.

Kunda, Ziva, and Richard E. Nisbett. "Prediction and the Partial Understanding of the Law of Large Numbers." *Journal of Experimental Social Psychology* 22 (1986): 339–54.

———. "The Psychometrics of Everyday Life." *Cognitive Psychology* 18 (1986): 195–224.

Kuo, Frances E., and William C. Sullivan. "Aggression and Violence in the Inner City: Effects of Environment via Mental Fatigue." *Environment and Behavior* 33 (2001): 543–71.

Lakatos, Imre. *The Methodology of Scientific Research Programmes*. Vol. 1, *Philosophical Papers*. Cambridge: Cambridge University Press, 1978.

Larrick, Richard P., J. N. Morgan, and R. E. Nisbett. "Teaching the Use of Cost-Benefit Reasoning in Everyday Life." *Psychological Science* 1 (1990): 362–70.

Larrick, Richard P., R. E. Nisbett, and J. N. Morgan. "Who Uses the Cost-Benefit Rules of Choice? Implications for the Normative Status of Microeconomic Theory." *Organizational Behavior and Human Decision Processes* 56 (1993): 331–47.

Lee, S.W.S., and N. Schwarz. "Bidirectionality, Mediation, and Moderation of Metaphorical Effects: The Embodiment of Social Suspicion and Fishy Smells." *Journal of Personality and Social Psychology* (2012). Published electronically August 20, 2012.

Lehman, Darrin R., Richard O. Lempert, and Richard E. Nisbett. "The Effects of Graduate Training on Reasoning: Formal Discipline and Thinking About Everyday Life Events." *American Psychologist* 43 (1988): 431–43.

Lehman, Darrin R., and Richard E. Nisbett. "A Longitudinal Study of the Effects of Undergraduate Education on Reasoning." *Developmental Psychology* 26 (1990): 952–60.

Lepper, Mark R., David Greene, and Richard E. Nisbett. "Undermining Children's Intrinsic Interest with Extrinsic Reward: A Test of the Overjustification Hypothesis." *Journal of Personality and Social Psychology* 28 (1973): 129–37.

Levin, Irwin P., and Gary J. Gaeth. "Framing of Attribute Information Before and After Consuming the Product." *Journal of Consumer Research* 15 (1988): 374–78.

Levitt, Steven D., and Stephen J. Dubner. *Freakonomics: A Rogue Economist Explores the Hidden Side of Everything.* New York: William Morrow, 2005.

Lewicki, Pawel, Thomas Hill, and Maria Czyzewska. "Nonconscious Acquisition of Information." *American Psychologist* 47 (1992): 796–801.

Lichtenfield, S., A. J. Elliot, M. A. Maier, and R. Pekrun. "Fertile Green: Green Facilitates Creative Performance." *Personality and Social Psychology Bulletin* 38 (2012): 784–97.

Liu, Amy, S. Ruiz, L. DeAngelo, and J. Pryor. "Findings from the 2008 Administration of the College Senior Survey (CSS): National Aggregates." Los Angeles: University of California, Los Angeles, 2009.

Logan, Robert K. *The Alphabet Effect.* New York: Morrow, 1986.

Lowry, Annie. "Caught in a Revolving Door of Unemployment." *The New York Times,* November 16, 2013.

Lu, J-C, F. Chen, H-R Xu, and N-Q Lu. "Comparison of Three Sperm-Counting Methods for the Determination of Sperm Concentration in Human Semen and Sperm Suspensions." *LabMedicine* 38 (2007): 232–36.

Madrian, Brigitte C., and Dennis F. Shea. "The Power of Suggestion: Inertia in 401(K) Participation and Savings Behavior." *Quarterly Journal of Economics* 116, no. 4 (2001): 1149–1225.

Magnuson, K., C. Ruhm, and J. Waldfogel. "How Much Is Too Much? The Influence of Preschool Centers on Children's Social and Cognitive Development." *Economics of Education Review* 26 (2007): 52–66.

Maier, N.R.F. "Reasoning in Humans II: The Solution of a Problem and Its Appearance in Consciousness." *Journal of Comparative Psychology* 12 (1931): 181–94.

Masuda, Takahiko, P. C. Ellsworth, B. Mesquita, J. Leu, and E. van de Veerdonk. "Placing the Face in Context: Cultural Differences in the Perception of Facial Emotion." *Journal of Personality and Social Psychology* 94 (2008): 365–81.

Masuda, Takahiko, and Richard E. Nisbett. "Attending Holistically Vs. Analytically: Comparing the Context Sensitivity of Japanese and Americans." *Journal of Personality and Social Psychology* 81 (2001): 922–34.

McDade, T. W., J. Rutherford, L. Adair, and C. W. Kuzawa. "Early Origins of Inflammation: Microbial Exposures in Infancy Predict Lower Levels of C-Reactive Protein in Adulthood." *Proceedings of the Royal Society B* 277 (2010): 1129–37.

McNeil, B. J., S. G. Pauker, H. C. Sox, and A. Tversky. "On the Elicitation of Preferences for Alternative Therapies." *New England Journal of Medicine* 306 (1982): 943–55.

McPhee, J. "Draft No. 4: Replacing the Words in Boxes." *The New Yorker*, April 29, 2013.

Mehta, R., and R. Zhu. "Blue or Red? Exploring the Effect of Color on Cognitive Task Performances." *Science* 323 (2009): 1226–29.

Meyer, David E., and R. W. Schvaneveldt. "Facilitation in Recognizing Pairs of Words: Evidence of a Dependence Between Retrieval Operations." *Journal of Experimental Psychology* 90 (1971): 227–34.

Milkman, Katherine L., Modupe Akinola, and Dolly Chugh. "Temporal Distance and Discrimination: An Audit Study in Academia." *Psychological Science* (2012): 710–17.

Morris, Michael W., and Richard E. Nisbett. "Tools of the Trade: Deductive Reasoning Schemas Taught in Psychology and Philosophy." In *Rules for Reasoning*, edited by Richard E. Nisbett. Hillsdale, NJ: Lawrence Erlbaum, 1993.

Moss, Michael. "Nudged to the Produce Aisle by a Look in the Mirror." *The New York Times.* Published electronically August 28, 2013. http://www.nytimes.com/2013/08/28/dining/wooing-us-down-the-produce-aisle.html.

Mullainathan, Sendhil, and Eldar Shafir. *Scarcity: Why Having Too Little Means So Much.* New York: Times Books, 2013.

Munk, Nina. *The Idealist.* New York: Doubleday, 2013.

Naumann, Laura P., and O. John. "Are Asian Americans Lower in Conscientiousness and Openness?" Unpublished manuscript, 2013.

NBC News. "How to Value Life? EPA Devalues Its Estimate." Published electronically July 10, 2008. http://www.nbcnews.com/id/25626294/ns/us_news-environment/t/how-value-life-epa-devalues-its-estimate/#.Ucn7ZW3Q5Zp.

Nisbett, Richard E. *The Geography of Thought: How Asians and Westerners Think Differently . . . and Why.* New York: The Free Press, 2003.

——. "Hunger, Obesity and the Ventromedial Hypothalamus." *Psychological Review* 79 (1972): 433–53.

——. *Intelligence and How to Get It: Why Schools and Cultures Count.* New York: W. W. Norton, 2009.

——. *Rules for Reasoning.* Hillsdale, NJ: Lawrence Erlbaum, 1993.

Nisbett, Richard E., C. Caputo, P. Legant, and J. Maracek. "Behavior as Seen by the Actor and as Seen by the Observer." *Journal of Personality and Social Psychology* 27 (1973): 154–64.

Nisbett, Richard E., Geoffrey T. Fong, Darrin R. Lehman, and P. W. Cheng. "Teaching Reasoning." *Science* 238 (1987): 625–31.

Nisbett, Richard E., David H. Krantz, Christopher Jepson, and Geoffrey T. Fong. "Improving Inductive Inference." In *Judgment Under Uncertainty: Heuristics and Biases*, edited by D. Kahneman, P. Slovic, and A. Tversky. New York: Cambridge University Press, 1982.

Nisbett, Richard E., David H. Krantz, C. Jepson, and Ziva Kunda. "The Use of Statistical Heuristics in Everyday Reasoning." *Psychological Review* 90 (1983): 339–63.

Nisbett, Richard E., K. Peng, I. Choi, and A. Norenzayan. "Culture and Systems of Thought: Holistic Vs. Analytic Cognition." *Psychological Review* 108 (2001): 291–310.

Nisbett, Richard E., and L. Ross. *Human Inference: Strategies and Shortcomings of Social Judgment*. Englewood Cliffs, NJ: Prentice-Hall, 1980.

Nisbett, Richard E., and Timothy De Camp Wilson. "Telling More Than We Can Know: Verbal Reports on Mental Processes." *Psychological Review* 84 (1977): 231–59.

Nock, Matthew K., J. M. Park, C. T. Finn, T. L. Deliberto, H. J. Dour, and M. R. Banaji. "Measuring the Suicidal Mind: Implicit Cognition Predicts Suicidal Behavior." *Psychological Science* (2010). Published electronically March 9, 2010. http://pss.sagepub.com/content/21/4/511.

Norenzayan, Ara, and B. J. Kim. "A Cross-Cultural Comparison of Regulatory Focus and Its Effect on the Logical Consistency of Beliefs." Unpublished manuscript, 2002.

Norenzayan, A., E. E. Smith, B. J. Kim, and R. E. Nisbett. "Cultural Preferences for Formal Versus Intuitive Reasoning." *Cognitive Science* 26 (2002): 653–84.

O'Brien, Barbara, Samuel R. Sommers, and Phoebe C. Ellsworth. "Ask and What Shall Ye Receive? A Guide for Using and Interpreting What Jurors Tell Us." Digital commons at Michigan State University College of Law (2011). Published electronically January 1, 2011. http://digitalcommons.law.msu.edu/cgi/viewcontent.cgi?article=1416&context=facpubs.

Offit, Paul A. *Do You Believe in Magic? The Sense and Nonsense of Alternative Medicine*. New York: Harper-Collins, 2013.

Olszak, Torsten, D. An, S. Zeissig, M. P. Vera, J. Richter, A. Franke, J. N. Glickman et al. "Microbial Exposure During Early Life Has Persistent Effects on Natural Killer T Cell Function." *Science* 336 (2012): 489–93.

Parker-Pope, Tara, "The Decisive Marriage." The Well Column, *The New York Times*, August 25, 2014. www.well.blogs.nytimes.com/2014/08/25/the-decisive-marriage/?_r=0.

Peng, Kaiping. "Naive Dialecticism and Its Effects on Reasoning and Judgment About Contradiction." PhD dissertation, University of Michigan, 1997.

Peng, Kaiping, and Richard E. Nisbett. "Culture, Dialectics, and Reasoning About Contradiction." *American Psychologist* 54 (1999): 741–54.

Peng, Kaiping, Richard E. Nisbett, and Nancy Y. C. Wong. "Validity Problems Comparing Values Across Cultures and Possible Solutions." *Psychological Methods* 2 (1997): 329–44.

Peng, Kaiping, Julie Spencer-Rodgers, and Zhong Nian. "Naive Dialecticism and the Tao of Chinese Thought." In *Indigenous and Cultural Psychology: Understanding People in Context*, edited by Uichol Kim, Kuo-Shu Yang, and Kwang-Kuo Hwang. New York: Springer, 2006.

Pennebaker, James W. "Putting Stress into Words: Health, Linguistic and Therapeutic Implications." *Behavioral Research and Therapy* 31 (1993): 539–48.

Perkins, H. Wesley, Michael P. Haines, and Richard Rice. "Misperceiving the College Drinking Norm and Related Problems: A Nationwide Study of Exposure to Prevention Information, Perceived Norms and Student Alcohol Misuse." *Journal of Studies on Alcohol* 66 (2005): 470–78.

Pietromonaco, Paula R., and Richard E. Nisbett. "Swimming Upstream Against the Fundamental Attribution Error: Subjects' Weak Generalizations from the Darley and Batson Study." *Social Behavior and Personality* 10 (1982): 1–4.

Polanyi, Michael. *Personal Knowledge: Toward a Post-Critical Philosophy.* New York: Harper & Row, 1958.

Prentice, Deborah A., and Dale T. Miller. "Pluralistic Ignorance and Alcohol Use on Campus: Some Consequences of Misperceiving the Social Norm." *Journal of Personality and Social Psychology* 64 (1993): 243–56.

Rein, Martin, and Lee Rainwater. "How Large Is the Welfare Class?" *Challenge* (September–October 1977): 20–33.

Riegel, Klaus F. "Dialectical Operations: The Final Period of Cognitive Development." *Human Development* 18 (1973): 430–43.

Rigdon, M., K. Ishii, M. Watabe, and S. Kitayama. "Minimal Social Cues in the Dictator Game." *Journal of Economic Psychology* 30 (2009): 358–67.

Roberts, N. P., N. J. Kitchiner, J. Kenardy, and J. Bisson. "Multiple Session Early Psychological Interventions for Prevention of Post-Traumatic Disorder." *Cochrane Summaries* (2010). http://summaries.cochrane.org/CD006869/multiple-session-early-psychological-interventions-for-prevention-of-post-traumatic-stress-disorder.

Ross, L., and A. Ward. "Naive Realism in Everyday Life: Implications for Social Conflict and Misunderstanding." In *Values and Knowledge*, edited by E. Reed, T. Brown, and E. Turiel. Hillsdale, NJ: Erlbaum, 1996.

Saad, Lydia. "U.S. Abortion Attitudes Closely Divided." Gallup Poll (2009). http://www.gallup.com/poll/122033/u.s.-abortion-attitudes-closely-divided.aspx.

Samieri, C., C. Feart, C. Proust-Lima, E. Peuchant, C. Tzourio, C. Stapf, C. Berr, and P. Barberger-Gateau. "Olive Oil Consumption, Plasma Oleic Acid, and Stroke Incidence." *Neurology* 77 (2011): 418–25.

Samuel, Lawrence R. *Shrink: A Cultural History of Psychoanalysis in America.* Lincoln, NE: University of Nebraska Press, 2013.

Samuelson, William, and Richard J. Zeckhauser. "Status Quo Bias in Decision Making." *Journal of Risk and Uncertainty* 1 (1988): 7–59.

Sanchez-Burks, Jeffrey. "Performance in Intercultural Interactions at Work: Cross-Cultural Differences in Responses to Behavioral Mirroring." *Journal of Applied Psychology* 94 (2009): 216–23.

Schmitt, David P., J. Allik, R. R. McCrae, and V. Benet-Martinez. "The Geographic Distribution of Big Five Personality Traits: Patterns and Profiles of Human Self-Description Across 56 Nations." *Journal of Cross-Cultural Psychology* 38 (2007): 173–212.

Schnall, E., S. Wassertheil-Smoller, C. Swencionis, V. Zemon, L. Tinker, J. O'Sullivan, et al. "The Relationship Between Religion and Cardiovascular Outcomes and All-Cause Mortality: The Women's Health Initiative Observational Study (Electronic Version)." *Psychology and Health* (2008): 1–15.

Schultz, P. Wesley, J. M. Nolan, R. B. Cialdini, N. J. Goldstein, and V. Griskevicius. "The Constructive, Destructive, and Reconstructive Power of Social Norms." *Psychological Science* 18 (2007): 429–34.

Schwarz, Norbert, and Gerald L. Clore. "Mood, Misattribution, and Judgments of Well-Being: Informative and Directive Functions of Affective States." *Journal of Personality and Social Psychology* 45 (1983): 513–23.

Schwarz, Norbert, Fritz Strack, and Hans-Peter Mai. "Assimilation-Contrast Effects in Part-Whole Question Sequences: A Conversational Logic Analysis." *Public Opinion Quarterly* 55 (1991): 3–23.

Seligman, Martin E. P. "On the Generality of the Laws of Learning." *Psychological Review* 77 (1970): 127–90.

Shepard, Roger N. *Mind Sights: Original Visual Illusions, Ambiguities, and Other Anomalies.* New York: W. H. Freeman and Company, 1990.

Shin, In-Soo, and Jae Young Chung. "Class Size and Student Achievement in the United States: A Meta-Analysis." *Korean Educational Institute Journal of Educational Policy* 6 (2009): 3–19.

Silver, Nate. *The Signal and the Noise.* New York: The Penguin Press, 2012.

Slomski, Anita. "Prophylactic Probiotic May Prevent Colic in Newborns." *Journal of the American Medical Association* 311 (2014).

Smedslund, Jan. "The Concept of Correlation in Adults." *Scandinavian Journal of Psychology* 4 (1963): 165–73.

Song, H., and N. Schwarz. "If It's Hard to Read, It's Hard to Do." *Psychological Science* 19 (2008): 986–88.

Stephens-Davidowitz, Seth. "Dr. Google Will See You Now." *The New York Times,* August 11, 2013.

Stich, Stephen, ed. *Collected Papers: Knowledge, Rationality, and Morality, 1978–2010.* New York: Oxford, 2012.

Strack, Fritz, Leonard L. Martin, and Sabine Stepper. "Inhibiting and Facilitating Conditions of the Human Smile: A Nonobtrusive Test of the Facial Feedback Hypothesis." *Journal of Personality and Social Psychology* 53 (1988): 768–77.

Straub, Kath. "Mind the Gap: On the Appropriate Use of Focus Groups and Usability Testing in Planning and Evaluating Interfaces." In *Human Factors International: Free Resources Newsletter,* September 2004.

Strick, M., A. Dijksterhuis, M. W. Bos, A. Sjoerdsma, and R. B. van Baaren. "A Meta-Analysis on Unconscious Thought Effects." *Social Cognition* 29 (2011): 738–62.

Sunstein, Cass R. "The Stunning Triumph of Cost-Benefit Analysis." *Bloomberg View* (2012). Published electronically September 12, 2012. http://www.bloomberg.com /news/2012-09-12/the-stunning-triumph-of-cost-benefit-analysis.html.

Thaler, Richard H., and C. R. Sunstein. *Nudge: Improving Decisions About Health, Wealth, and Happiness.* New York: Penguin Books, 2008.

Thompson, William C., Geoffrey T. Fong, and D. L. Rosenhan. "Inadmissible Evidence and Juror Verdicts." *Journal of Personality and Social Psychology* 40 (1981): 453–63.

Triplett, Norman. "The Dynamogenic Factors in Pacemaking and Competition." *American Journal of Psychology* 9 (1898): 507–33.

Tversky, Amos, and Daniel Kahneman. "Extensional Versus Intuitive Reasoning: The Conjunction Fallacy in Probability Judgment." *Psychological Review* 90 (1983): 293–315.

———. "Judgment Under Uncertainty: Heuristics and Biases." *Science* 185 (1974): 1124–31.

Valochovic, R. W., C. W. Douglass, C. S. Berkey, B. J. McNeil, and H. H. Chauncey. "Examiner Reliability in Dental Radiography." *Journal of Dental Research* 65 (1986): 432–36.

Ward, W. D., and H. M. Jenkins. "The Display of Information and the Judgment of Contingency." *Canadian Journal of Psychology* 19 (1965): 231–41.

Watanabe, M. "Styles of Reasoning in Japan and the United States: Logic of Education in Two Cultures." Paper presented at the American Sociological Association, San Francisco, CA, 1998.

Weiss, J., and P. Brown. "Self-Insight Error in the Explanation of Mood." Unpublished manuscript, 1977.

Williams, Lawrence E., and John A. Bargh. "Experiencing Physical Warmth Influences Personal Warmth." *Science* 322 (2008): 606–607.

Wilson, Timothy D. *Redirect: The Surprising New Science of Psychological Change.* New York: Little, Brown, 2011.

Wilson, T. D., and J. W. Schooler. "Thinking Too Much: Introspection Can Reduce the Quality of Preferences and Decisions." *Journal of Personality and Social Psychology* 60 (1991): 181–92.

Wolf, Pamela H., J. H. Madans, F. F. Finucane, and J. C. Kleinman. "Reduction of Cardiovascular Disease–Related Mortality Among Postmenopausal Women Who Use Hormones: Evidence from a National Cohort." *American Journal of Obstetrics and Gynecology* 164 (1991): 489–94.

Zagorsky, Jay L. "Do You Have to Be Smart to Be Rich? The Impact of IQ on Wealth, Income and Financial Distress." *Intelligence* 35 (2007): 489–501.

Zajonc, Robert B. "The Attitudinal Effects of Mere Exposure." *Journal of Personality and Social Psychology* 9 (1968): 1–27.

Zebrowitz, Leslie. *Reading Faces: Window to the Soul?* Boulder, CO: Westview Press, 1997.

Acknowledgments

Many people gave me valuable criticism and advice in the writing of this book. They include Ray Batra, Sara Billmann, Dov Cohen, Christopher Dahl, William Dickens, Phoebe Ellsworth, James Flynn, Thomas Gilovich, Igor Grossmann, Keith Holyoak, Gordon Kane, Shinobu Kitayama, Darrin Lehman, Michael Maharry, Michael Morris, Lee Ross, Justin Sarkis, Norbert Schwarz, Stephen Stich, Carol Tavris, Paul Thagard, Amiram Vinokur, Kenneth Warner, and Timothy Wilson. I feel very lucky to have John Brockman and Katinka Matson as my literary agents.

I am deeply indebted to my sage editor, Eric Chinski, who functioned like a valued colleague. Peng Shepherd and the rest of the editorial staff at Farrar, Straus and Giroux were extremely helpful and patient.

Susan Nisbett made the book better in every way, from discussion of ideas to editing. She also makes my life better in every way.

I owe a great deal to the University of Michigan, whose environment encourages interdisciplinary research. Many fields of scientific research have been created there at the intersection of older disciplines. Participating in this ferment has helped me to see the extent to which science is a seamless web.

Index

Page numbers in *italics* refer to illustrations.

ALLEN LANE
an imprint of
PENGUIN BOOKS

Recently Published

David Wootton, *The Invention of Science: A New History of the Scientific Revolution*

Christopher Tyerman, *How to Plan a Crusade: Reason and Religious War in the Middle Ages*

Andy Beckett, *Promised You A Miracle: UK 80–82*

Carl Watkins, *Stephen: The Reign of Anarchy*

Anne Curry, *Henry V: From Playboy Prince to Warrior King*

John Gillingham, *William II: The Red King*

Roger Knight, *William IV: A King at Sea*

Douglas Hurd, *Elizabeth II: The Steadfast*

Richard Nisbett, *Mindware: Tools for Smart Thinking*

Jochen Bleicken, *Augustus: The Biography*

Paul Mason, *PostCapitalism: A Guide to Our Future*

Frank Wilczek, *A Beautiful Question: Finding Nature's Deep Design*

Roberto Saviano, *Zero Zero Zero*

Owen Hatherley, *Landscapes of Communism: A History Through Buildings*

César Hidalgo, *Why Information Grows: The Evolution of Order, from Atoms to Economies*

Aziz Ansari and Eric Klinenberg, *Modern Romance: An Investigation*

Sudhir Hazareesingh, *How the French Think: An Affectionate Portrait of an Intellectual People*

Steven D. Levitt and Stephen J. Dubner, *When to Rob a Bank: A Rogue Economist's Guide to the World*

Leonard Mlodinow, *The Upright Thinkers: The Human Journey from Living in Trees to Understanding the Cosmos*

Hans Ulrich Obrist, *Lives of the Artists, Lives of the Architects*

Richard H. Thaler, *Misbehaving: The Making of Behavioural Economics*

Sheldon Solomon, Jeff Greenberg and Tom Pyszczynski, *Worm at the Core: On the Role of Death in Life*

Nathaniel Popper, *Digital Gold: The Untold Story of Bitcoin*

Dominic Lieven, *Towards the Flame: Empire, War and the End of Tsarist Russia*

Noel Malcolm, *Agents of Empire: Knights, Corsairs, Jesuits and Spies in the Sixteenth-Century Mediterranean World*

James Rebanks, *The Shepherd's Life: A Tale of the Lake District*

David Brooks, *The Road to Character*

Joseph Stiglitz, *The Great Divide*

Ken Robinson and Lou Aronica, *Creative Schools: Revolutionizing Education from the Ground Up*

Clotaire Rapaille and Andrés Roemer, *Move UP: Why Some Cultures Advances While Others Don't*

Jonathan Keates, *William III and Mary II: Partners in Revolution*

David Womersley, *James II: The Last Catholic King*

Richard Barber, *Henry II: A Prince Among Princes*

Jane Ridley, *Victoria: Queen, Matriarch, Empress*

John Gray, *The Soul of the Marionette: A Short Enquiry into Human Freedom*

Emily Wilson, *Seneca: A Life*

Michael Barber, *How to Run a Government: So That Citizens Benefit and Taxpayers Don't Go Crazy*

Dana Thomas, *Gods and Kings: The Rise and Fall of Alexander McQueen and John Galliano*

Steven Weinberg, *To Explain the World: The Discovery of Modern Science*

Jennifer Jacquet, *Is Shame Necessary?: New Uses for an Old Tool*

Eugene Rogan, *The Fall of the Ottomans: The Great War in the Middle East, 1914-1920*

Norman Doidge, *The Brain's Way of Healing: Stories of Remarkable Recoveries and Discoveries*

John Hooper, *The Italians*

Sven Beckert, *Empire of Cotton: A New History of Global Capitalism*

Mark Kishlansky, *Charles I: An Abbreviated Life*

Philip Ziegler, *George VI: The Dutiful King*

David Cannadine, *George V: The Unexpected King*

Stephen Alford, *Edward VI: The Last Boy King*

John Guy, *Henry VIII: The Quest for Fame*

Robert Tombs, *The English and their History: The First Thirteen Centuries*

Neil MacGregor, *Germany: The Memories of a Nation*

Uwe Tellkamp, *The Tower: A Novel*

Roberto Calasso, *Ardor*

Slavoj Žižek, *Trouble in Paradise: Communism After the End of History*

Francis Pryor, *Home: A Time Traveller's Tales from Britain's Prehistory*

R. F. Foster, *Vivid Faces: The Revolutionary Generation in Ireland, 1890-1923*

Andrew Roberts, *Napoleon the Great*

Shami Chakrabarti, *On Liberty*

Bessel van der Kolk, *The Body Keeps the Score: Mind, Brain and Body in the Transformation of Trauma*

Brendan Simms, *The Longest Afternoon: The 400 Men Who Decided the Battle of Waterloo*

Naomi Klein, *This Changes Everything: Capitalism vs the Climate*

Owen Jones, *The Establishment: And How They Get Away with It*

Caleb Scharf, *The Copernicus Complex: Our Cosmic Significance in a Universe of Planets and Probabilities*

Martin Wolf, *The Shifts and the Shocks: What We've Learned - and Have Still to Learn - from the Financial Crisis*

Steven Pinker, *The Sense of Style: The Thinking Person's Guide to Writing in the 21st Century*

Vincent Deary, *How We Are: Book One of the How to Live Trilogy*

Henry Kissinger, *World Order*

Alexander Watson, *Ring of Steel: Germany and Austria-Hungary at War, 1914-1918*

Richard Vinen, *National Service: Conscription in Britain, 1945-1963*

Paul Dolan, *Happiness by Design: Finding Pleasure and Purpose in Everyday Life*

Mark Greengrass, *Christendom Destroyed: Europe 1517-1650*

Hugh Thomas, *World Without End: The Global Empire of Philip II*

Richard Layard and David M. Clark, *Thrive: The Power of Evidence-Based Psychological Therapies*

Uwe Tellkamp, *The Tower: A Novel*

Zelda la Grange, *Good Morning, Mr Mandela*

Ahron Bregman, *Cursed Victory: A History of Israel and the Occupied Territories*

Tristram Hunt, *Ten Cities that Made an Empire*

Jordan Ellenberg, *How Not to Be Wrong: The Power of Mathematical Thinking*

David Marquand, *Mammon's Kingdom: An Essay on Britain, Now*

Justin Marozzi, *Baghdad: City of Peace, City of Blood*

Adam Tooze, *The Deluge: The Great War and the Remaking of Global Order 1916-1931*

John Micklethwait and Adrian Wooldridge, *The Fourth Revolution: The Global Race to Reinvent the State*

Steven D. Levitt and Stephen J. Dubner, *Think Like a Freak: How to Solve Problems, Win Fights and Be a Slightly Better Person*

Alexander Monro, *The Paper Trail: An Unexpected History of the World's Greatest Invention*

Jacob Soll, *The Reckoning: Financial Accountability and the Making and Breaking of Nations*

Gerd Gigerenzer, *Risk Savvy: How to Make Good Decisions*

James Lovelock, *A Rough Ride to the Future*

Michael Lewis, *Flash Boys*

Hans Ulrich Obrist, *Ways of Curating*

Mai Jia, *Decoded: A Novel*

Richard Mabey, *Dreams of the Good Life: The Life of Flora Thompson and the Creation of* Lark Rise to Candleford

Danny Dorling, *All That Is Solid: The Great Housing Disaster*

Leonard Susskind and Art Friedman, *Quantum Mechanics: The Theoretical Minimum*

Michio Kaku, *The Future of the Mind: The Scientific Quest to Understand, Enhance and Empower the Mind*

Nicholas Epley, *Mindwise: How we Understand what others Think, Believe, Feel and Want*

Geoff Dyer, *Contest of the Century: The New Era of Competition with China*

Yaron Matras, *I Met Lucky People: The Story of the Romani Gypsies*

Larry Siedentop, *Inventing the Individual: The Origins of Western Liberalism*

Dick Swaab, *We Are Our Brains: A Neurobiography of the Brain, from the Womb to Alzheimer's*

Max Tegmark, *Our Mathematical Universe: My Quest for the Ultimate Nature of Reality*

David Pilling, *Bending Adversity: Japan and the Art of Survival*

Hooman Majd, *The Ministry of Guidance Invites You to Not Stay: An American Family in Iran*

Roger Knight, *Britain Against Napoleon: The Organisation of Victory, 1793-1815*

Alan Greenspan, *The Map and the Territory: Risk, Human Nature and the Future of Forecasting*

Daniel Lieberman, *Story of the Human Body: Evolution, Health and Disease*

Malcolm Gladwell, *David and Goliath: Underdogs, Misfits and the Art of Battling Giants*

Paul Collier, *Exodus: Immigration and Multiculturalism in the 21st Century*

John Eliot Gardiner, *Music in the Castle of Heaven: Immigration and Multiculturalism in the 21st Century*

Catherine Merridale, *Red Fortress: The Secret Heart of Russia's History*

Ramachandra Guha, *Gandhi Before India*

Vic Gatrell, *The First Bohemians: Life and Art in London's Golden Age*

Richard Overy, *The Bombing War: Europe 1939-1945*

Charles Townshend, *The Republic: The Fight for Irish Independence, 1918-1923*

Eric Schlosser, *Command and Control*

Sudhir Venkatesh, *Floating City: Hustlers, Strivers, Dealers, Call Girls and Other Lives in Illicit New York*

Sendhil Mullainathan and Eldar Shafir, *Scarcity: Why Having Too Little Means So Much*

John Drury, *Music at Midnight: The Life and Poetry of George Herbert*

Philip Coggan, *The Last Vote: The Threats to Western Democracy*

Richard Barber, *Edward III and the Triumph of England*

Daniel M Davis, *The Compatibility Gene*

John Bradshaw, *Cat Sense: The Feline Enigma Revealed*

Roger Knight, *Britain Against Napoleon: The Organisation of Victory, 1793-1815*

Thurston Clarke, *JFK's Last Hundred Days: An Intimate Portrait of a Great President*

Jean Drèze and Amartya Sen, *An Uncertain Glory: India and its Contradictions*

Rana Mitter, *China's War with Japan, 1937-1945: The Struggle for Survival*

Tom Burns, *Our Necessary Shadow: The Nature and Meaning of Psychiatry*